"Unless you follow politics closely, you probably don't know too much about Michael Deaver or how important he was to Ronnie as Governor of California, President of the United States, and as a friend. I think every president can point to one particular aide who knows him so well that he becomes indispensable. Mike was that man for Ronald Reagan, and there are few people better suited to write about him.

"He is the only one who was there from 1967 to 1987 with only the briefest of breaks, and at Ronnie's side for all the good times, and, of course, many of the tough times, too.

"For more than twenty years, Mike spent more time with Ronnie than anybody outside his family. Often it was only Mike at Ronnie's side in the Oval Office, on the campaign planes, and in motorcades. It is many of these memories that he shares in *A DIFFERENT DRUMMER*."

—from the *Foreword* by Nancy Reagan

Also by Michael K. Deaver

Nancy
Behind the Scenes

A DIFFERENT DRUMMER

MY THIRTY YEARS WITH
RONALD REAGAN

MICHAEL K. DEAVER
WITH A FOREWORD BY NANCY REAGAN

HarperTorch
An Imprint of HarperCollinsPublishers

HARPERTORCH
An Imprint of HarperCollins*Publishers*
10 East 53rd Street
New York, New York 10022-5299

First HarperTorch paperback printing: July 2004
First HarperCollins hardcover printing: May 2001

HarperCollins®, HarperTorch™, and ❦™ are trademarks of HarperCollins Publishers Inc.

Printed in the United States of America

Visit HarperTorch on the World Wide Web at www.harpercollins.com

10 9 8 7 6 5 4 3 2 1

*To Carolyn, for her counsel, love,
and the best of my days.*

Contents

Foreword

By Nancy Reagan

Unless you follow politics closely, you probably don't know too much about Michael Deaver or how important he was to Ronnie as governor of California, president of the United States, and as a friend. I think every president can point to one particular aide who knows him so well that he becomes indispensable. Mike was that man for Ronald Reagan, and there are few people better suited to write about him.

Since Ronnie left elective office, there have been many books, both pro and con, about his public life. Some have delved with limited success into a discussion of Reagan the individual.

I've always said that, with Ronnie, what you see is what you get. He is not a complicated man, and those who know him well won't disagree. Most tend either to overcomplicate him, or they err on the other side, wrongly labeling him as aloof, indifferent, or disengaged. Of course, none of these characterizations rings true.

Much of what has been written was written out of frustration brought about by the writers' inability to get at the man behind the politics, the presidency, and the public persona. They know the public Ronald Reagan; they do not know the private man. This is not surprising, because when you come right down to it, there are only a handful of people who could write about the private Ronald Reagan.

Without demeaning the other close relationships Ronnie has had over the years, undoubtedly one of the most qualified people to write about him in this regard is our friend Mike Deaver. Mike has been part of our lives since 1967 when he joined the governor's staff in Sacramento. He always occupied a unique role in Ronnie's life. He was never an official policy adviser and didn't serve as chief of staff, either in Washington or Sacramento. But he is the only one who was there from 1967 to 1987 with only the briefest of breaks, at Ronnie's side for all the good times, and of course, many of the tough times, too.

Mike and I have always shared a common interest. Our goal was simple: to help Ronnie be the best possible governor and president. What I came to love about Mike was that he was never the type to go on television and talk about his accomplishments. Never prone to self-promotion, he did his best work behind the scenes.

For more than twenty years, Mike spent more time with Ronnie than anybody outside his family. Often it was only Mike at Ronnie's side in the Oval Office, on the campaign planes, and in motorcades. It is many of these memories that he shares in *A Different Drummer.*

A DIFFERENT
DRUMMER

Introduction

At a time when much of the United States was in doubt and confusion, Ronald Reagan made his grand entrance onto the national stage with remarkable confidence, not so much in himself as in America. His unwavering optimism changed the way we thought about ourselves and the way the world looked at us. As I write these pages, America embarks on a new war, a war not unlike the battle Reagan led against the Soviet Union. We again face a committed enemy, one that may take decades to vanquish.

Like Reagan, George W. Bush has made the conflict the focus of his presidency, and I am confident that future historians will make many comparisons to the two leaders' leadership in mobilizing the American people for a long, protracted struggle against a gray, shadowy adversary.

As Bush begins his calling, historians are now chronicling the end of the Cold War so that future generations will understand Reagan's role in the demise of communism. I have my prejudices on the matter, but what I do know is that it is sad beyond expression that he can't acknowledge the many glorious achievements of his public life. Reagan lives today in a state of confusion, victimized by an illness that robs him of any coherent memory.

Perhaps because of this great, tragic irony, I felt a need to remember him, maybe even a duty. After all, I was lucky enough to have a front-row seat to this critical turning point in our country. I offer this book, then, as an attempt to place on the record my recollections of a remarkable man whom I had the pleasure of working alongside for more than twenty years.

A second deciding factor in writing *A Different Drummer* was Edmund Morris's much anticipated biography of Reagan, *Dutch*. Long before Morris had finished the book, it had become clear that he had struggled, and in his own mind failed, to truly understand Reagan. Lou Cannon, an esteemed journalist who has known Reagan far longer than Morris and has written extensively about him, once concluded it was impossible to figure him out entirely. The truth is that most people are more simple or complicated than can be fully explained by anyone, and Reagan is no exception. Still, I hope that I can make a small contribution to understanding Reagan by passing on these accounts of our time together. People can draw whatever conclusions they want from these anecdotes. I would never pretend to have solved the riddle of Reagan, and this book will hardly satisfy those who want a penetrating insight into the mystery of the man. Maybe, though, readers will find it entertaining and even enlightening at times.

I should pause here to tell you what this

book is not. This is not a biography in the sense of a work based on vast research and long months spent combing through personal archives. I have never kept a diary, nor do I have the inclination to pore over memoranda and public documents that summarize activities of that time. There will be no footnotes or excerpts from lengthy interviews. This is a book of memories inspired by a need to share whatever insights I can into a very public yet thoroughly private man who has confounded everyone who has tried to sum up his life.

The title of the book comes from my belief that Reagan was, in part, so difficult to understand because he was guided by a source of inspiration that only he understood. I have never known a person to be so sure of what he thought he knew or so accepting of things he could never understand. He was not a Hamlet, not a man of angst. Reagan had a profound spiritual faith that grounded him and left him with a nearly perpetual peace of mind. This is not to say he didn't have his moments of doubt and anger, but those times were rare. He didn't have trouble conveying confidence because he was sure of what he thought. He didn't have to fake it, his acting ability aside. His steady sense of purpose grew from his complete acquiescence to a drummer only he could hear.

As for his well-documented remoteness, the part of his personality that drives his biographers crazy, a lot of theories get bandied about as to what shaped Reagan's personality. He couldn't see very well as a child until his nearsightedness finally and belatedly led to him getting glasses. Is that why he was so disconnected from people? His father's drunkenness is often cited as having had a profound impact on his behavior. Perhaps this is the reason he learned to keep his distance from people, or why he invented a world of "make-believe" where America was always a beacon of hope on a shining hill. Far be it from me to analyze him. After all our years together, I could never completely understand what made him tick. I did, however, observe the traits that made him so successful as a leader and so peculiar—and wonderful—as a person.

He had phenomenal discipline in almost everything he did. Sure, he could go drinking with the boys, but never more than one drink. Much has been made of Reagan's so-called laziness, but the fact is he was the one who was plowing his way through briefing papers on the plane when the rest of us were too tired to think straight anymore. Night after night, day after day, he could deliver the same speech, knowing that as old as it was to everyone who traveled with him regularly, it was always new to those in the audience who were hearing it for the first time. He had amazing stamina, maintaining an exercise routine more rigorous than that of many people fifty years his junior. While many of us long for vacations spent reading, napping, and lying on a beach, his idea of relaxation was riding a horse all day, fixing fence posts, and chopping wood until his muscles ached and his hands blistered. Ronald Reagan worked eight years doing the toughest job on earth, crisscrossed the world, and survived an assassin's bullet, a devastating riding accident, cancer, and brain surgery all *after* he turned seventy.

While some in politics make lists of enemies, Reagan was forever bringing his foes "under the tent." He never used poll numbers to make up his mind. Whenever we showed him statistical evidence that the majority opposed his views, he told us we all had to work harder to convince people he was right. When I first met him, I didn't even think he could win the Republican nomination for governor of California. Somewhere between two huge gubernatorial victories and two landslide presidential elections, I realized no one understood the American people and voters better than Ronald Reagan. He spent a lifetime surprising people who underestimated him.

Sometimes his simple belief that Americans can become whatever they aspire to be absolutely drove me crazy. Then I would remember where he came from and how inexplicable it is that someone like Reagan could become president. And I began to understand why he believed everyone could become someone. His own political views were first shaped by the deep poverty of the Great Depression and a great respect for Franklin D. Roosevelt. As his political beliefs were transformed, he became a captivating speaker with a tremendous capacity for instilling great loyalty in his followers. Just as his political idol had lifted people out of despair with a powerful voice and a message of hope, Reagan, who would become known as the great communicator of his time, would also leave a legacy of inspiration.

One thing I hope this book helps clear up is the myth that Reagan's popularity was not of his own making—created instead by professional handlers. Those pundits who had a hard time understanding his appeal insisted on giving credit to his staff for making him so attractive to the American public. I was one of those staffers, and I can assure you it was Reagan who had the gift for connecting with people, not me or any of the rest of us. Whatever help we provided was marginal at best. It was Reagan's convictions, decisions, and personal magnetism that made it all work so well. We did not make Ronald Reagan. He's the one who made us.

His personal life was as peaceful as his professional life was exciting. If he never had to go to another party or fund-raiser, that would have been fine with him. When he did go to events, as all politicians do with unnerving frequency, he was painfully shy and much preferred talking with ordinary people over those with so-called star appeal. He was perfectly content slipping into pajamas in the evening and reading books or watching old movies with Nancy.

It is impossible for me to think of Ronald Reagan without thinking of Nancy. Their love story is the greatest one I have ever known and an inspiration to people everywhere. When I first began working for then Governor Reagan, one of my jobs was to work closely with Nancy. Because their marriage was so close, because they understood each other so deeply and completely, my work offered an extraordinary opportunity to get to know them both and to come to realize how complete they were together. As long as they had each other, they didn't really need anyone else. That may explain as much as anything else Reagan's ability to enjoy life as much as he did without close, personal friends. She was his fulfillment, and still is. As troubling as it is to watch such a wonderful marriage fall victim to Alzheimer's disease, that thief of memories, it's still comforting to know that Nancy is there for him, constantly at his side. It is clearly her finest hour.

When I told my friend Senator Connie Mack that I was writing this book, he immediately jumped out of his chair, handed me a white paper he had just completed, and directed me to a dog-eared page that boasted the title "The Reagan Legacy." I know what Reagan means to me, but I am always moved when I hear others try to place him in the context of their own lives and of history. Connie's words about what Reagan means to him and America are as eloquent as anything I have read:

Our debt to Ronald Reagan reminds me of an exchange mission I once went on, with Congressmen Tom Foley and Dick Cheney, among others. In 1985, a delegation from Congress went to France to reciprocate an earlier visit by a group of French legislators. After spending some time in Paris, each of us was assigned to a member of the National Assembly, and we accompanied them to their districts. I went to a district near Le Mans. An extensive tour ended in a town hall meeting, where some of my host's constituents questioned me for some two hours. At the end of the meeting, I asked my host, François, if I could ask them a question. He assented and I asked what they thought about America, so I could tell it to the people of Florida when I went home the next day.

The first person stood up and said, "We think of America as a dynamic, growing, thriving exciting place." A second man got up and said basically the same thing. The third person to address me was a man who was probably in his late seventies or early eighties. He was stooped over, his weight being supported precariously by an old gnarled cane. He came over closer to me, looked me directly in the eye, and said, "You tell the people of America that we will never forget that it was the American G.I. who saved our little town. You tell them we'll never forget."

Well, I feel that way about Ronald Reagan, my political hero, who inspired me in my very first run for office. America will never forget what President Reagan did for us. He gave us back faith and renewed our belief in this country. He gave America back its pride. He rebuilt America's defenses. His economic policies reduced taxes, reduced inflation, and reduced unemployment. He put America back to work again. He reminded America what made us a great nation. And he won the Cold War without firing one shot.

The citizens of America and the people of the world will never forget.

Nor will I. This book is my way of helping me and all of us remember this remarkable man.

CHAPTER ONE

THE EARLY YEARS

I stood alone in the comfortable office Ronald Reagan had used since leaving the White House. He had selected this particular office, I think, for its unforgettable view of the Pacific Ocean. This is the panorama every Southern Californian dreams of: the winding coast from Malibu to Long Beach, covered in an early-morning fog that breaks around noon most days and ends with a stunning sunset over the water. Except for his years in Sacramento and Washington, D.C., this was the view that Reagan had known and loved all his adult life. His home in Bel Air, his office at the Ronald Reagan Library in Simi Valley, the getaway at Rancho

De Cielo in the Santa Barbara Hills all opened up to essentially the same scene. His and Nancy's final resting place, a tomb built for the two of them at the library, had been chosen in part because it looked across the Ventura coast to Point Mugu.

The office I was standing in was all Reagan as well: the chair he used as governor, a painting behind the desk by Charles Reiffle titled *Desert Below Julian*, other paintings with western themes including several of his ranch and one of two of his horses, reproductions of Frederick Remington's sculptures of cowboys, Indians, and horses. There were photos of his four children, too, and of Nancy, and one of himself with Dwight Eisenhower that he has had in every office since I've known him. As I stood there, my eye fell on the little glass dome in one of the cabinets. I didn't have to move closer to know what was inside: a .22-caliber bullet inscribed "This Is It!" along with a shard of glass from the limousine Reagan had been about to climb into on that awful day in 1981 when John Hinckley had nearly ended his presidency and his life.

Today was an overcast Southern California morning in 1997, about three years after Ronald Reagan had penned his open letter disclosing his Alzheimer's disease, essentially saying his farewell to the American people. The letter, simple and touching, was written by a man who never claimed to be John Keats. It tells future generations in just a few paragraphs more about the man than any Reagan biography could.

As soon as Reagan finished that letter—as if on cue—the Alzheimer's seemed to worsen, deepen. He would never be the same after he put the pen down, and America would never see Citizen Reagan again. Ronald and Nancy Reagan had agreed there would be no more public appearances. The letter was his subtle yet eloquent exit from the grand stage.

This would be our first meeting since the letter was written. I felt fairly certain, too, that this would be the last time I would ever see Reagan alive. It wasn't—I would see him once more—but sooner or later, and probably sooner, Nancy was going to cut off access to him altogether, even for old friends like me.

She wanted him to be remembered as strong and energetic. Having spent so many years as, in effect, the cokeeper of that image, I could only sympathize. Over the course of the thirty years I had known him, I had come to love and respect Reagan like a second father. He was once asked if he thought of me as another son. He thought it over for a moment and said, "Son, no. Brother, maybe."

Joanne Drake, his chief of staff since leaving the White House, assured me that I would have to wait only a few minutes more. The president was on his way. I was actually nervous, wondering how he'd react. Would he remember me? What would I say to someone I'd worked side by side with for all those years? Someone whom I have seen at his highest heights and lowest depths?

What do you say to a man who, thanks to Alzheimer's, you know more about than he does?

There were a number of framed pictures on the shelves, which served as personal reminders for a man with a fading memory.

Joanne had told me they helped him link newly designated strangers like me with the past. Reagan had developed a routine he was very comfortable with when greeting visitors. He would walk them around the office, talking about specific photos or other objects so he could gauge the visitor's interest in each and begin to connect his visitor with his past. I could envision the drill. And I understood the gimmick: It was vintage Reagan. Take the focus off him and toss it casually upon your shoulders. With Reagan, it was never about himself.

The photos and mementos came in handy, as there were many like me who came calling, seeking just one more moment of camaraderie and friendship. That was my sole reason for being here this day. I was in Los Angeles and I wanted to see my old friend again.

The sound of stirrings in the outside office snapped me out of my reverie. Through the open door came Ronald Reagan. The man I called governor and Mr. President—only once would I call him Ron—was back in my life. A crooked smile creased his face as he extended his hand and took mine. I felt that old infectious joy and optimism he always seemed to carry into a room. His chestnut hair had given up the battle against time and was brushed with a dignified gray. I know for certain that no dye ever touched Reagan's hair. For years, the Reagan haters had literally sifted through his barber's trash can, searching for a dyed gray lock that could serve as a tiny metaphor for a phony man and an even more phony presidency. They searched in vain. It was an old actor's trick—Brylcreem—that gave Reagan's hair that dark gloss, not Clairol for Men.

His shoulders were a little stooped on this day, and his movements were slower than I had remembered. He looked docile and a bit worn. The man who had brought the Soviet bear to its knees now seemed more like a gentleman you might see in a Leisure World lobby, not the West Wing. Still, the look was the same. A crisp white shirt, a flawless dark suit with a half inch of cuff showing at the wrist.

We made small talk for a few minutes before

Reagan escorted me toward the pictures that documented his life, but I don't think he gleaned any clues of who I was based on my reactions. Undaunted, he fixed his eyes on the now-famous photo of himself as a young lifeguard at Rock River. He pointed to the hulking image with pride, asking if I had any idea how many lives he had saved during the time he kept watch there. Without hesitation, I belted out the precise number, "Seventy-seven."

Astonished, Reagan looked at me and demanded that I cite the source of my knowledge. It was like I had just revealed his most cherished secret. I wanted to remind him of the time, after he had left office, when we went to Rock River in Illinois and he showed me where he had pulled most of the people out of the water. An undercurrent ran through the spot. Swimmers would get caught up in it and panic, and then he would have to race out and overpower them, using his strength and his knowledge of the river to save their lives. He loved his lifeguarding days—they were a parable of his larger life as he saw it. But there was no sense saying any of that now. The time for sharing old stories had passed.

There were a few moments of silence as Reagan went back to examining his Great Wall of support of about ten photos for some sort of connection to Mike Deaver. Then the phone on his desk began chirping, and I think we were both thankful for the distraction. He politely excused himself, turning on his heels toward the desk. His gracefulness—unlike his memory—had not fallen victim to disease. Perhaps it was wishful thinking on my part, but as he strolled across the room, he seemed to get some of his old mojo, that famous Reagan stride, back. His posture straightened, the chest expanded. Even at his age he moved with an elegance that was as natural as breathing was to somebody like me.

He hung up the phone and, smiling, informed me that his lunch was ready as Joanne Drake walked in with a camera.

"Mike, how 'bout a photo of you two guys for old time's sake?" she asked.

Reagan didn't look at her, but walked directly toward me and grabbed my hand with both of his. There is no way to tell for sure, but as he pulled me in, I was hoping that he was thinking, "Oh yeah, it's Mike."

Our short meeting was over. As I drove back to the hotel, I found myself thinking about the first time I'd gripped that hand, thirty-five years earlier. In the years to come, we would travel the world together on Air Force One, but that first face-to-face hadn't been far from where I was just then—seventy miles up the coast, at a Santa Barbara restaurant called Talk of the Town. Maybe, I thought, Ronald Reagan didn't know any more about me now than he did then, but I'd gone to school on the man. And unlike him and the tough hand fate had dealt him, I had forgotten almost nothing about the experience.

I had seen Reagan for the first time a year earlier, in 1964, in the lobby of the Ambassador Hotel in downtown Los Angeles. I was just another faceless twentysomething back in those days, working quietly within the bowels of the

California Republican Party while we all tried to recover from Barry Goldwater's recent slaughter at the polls. I had been lured into the party by the Arizona senator's straight talk, his promise to do the right thing and implement a real conservative agenda.

The Ambassador event was just another in a seemingly endless stream of political fund-raisers, speeches, and grip-and-grin sessions; but the lobby was full. The doors to my left swung open and Ronald Wilson Reagan walked into the lobby unescorted. He glanced both ways, apparently seeking the person he was to meet. It was understandably tough to find anybody in this crowd, essentially white guys with ties and bellies too large for their overworked belts.

For me, Reagan instantly stood out, but nobody else seemed to notice him as he marched through the lobby toward the elevators. Later, there would be Secret Service, fawning aides, a chattering press corps, and television cameras in tow; but at this early date it was just Reagan, and truth told, I wasn't that impressed.

I had grown up stargazing in Southern California, I knew my stuff. I could pick out an actor when I saw one, and I was seeing one now. This guy looked like Hollywood incarnate. Everything was too perfect: the flawless hair, that robust chest, those rosy cheeks. I was sure he was wearing makeup. His walk was practiced and perfected—the only thing missing was the book on his head. Of course, I'd heard about this Ronald Reagan fellow. I knew his movies, vaguely, and I'd seen him hosting *Death Valley Days* and *GE Theater*. Word was that he could deliver some good zingers on the rubber-chicken circuit. In my book, none of that added up to much of a political future, but I must say, looking back over all those years, that there was an aura about him even then. And somebody, after all, would have to lead the party out of the ruins of the Goldwater debacle.

Nationwide, Lyndon Johnson's landslide victory had sharpened still further the natural divisions within the Republican Party: easterners versus westerners; moderates versus mainstream; right-wing "John Birchers" versus liberal Republicans. It was even more fractured in California, where everything, including ideology, gets taken to extremes. From these ashes, Reagan would rise.

The people who supported Goldwater in California already knew Reagan and suspected he could be the party's savior, but outside these circles, he would have to make his case that he was the man to bring all the factions together. On this day in the mid-1960s I would see for the first time the man who many California conservatives said had the potential to do what Barry Goldwater couldn't do—win. Soon, I'd begin to get a front-row view.

The next year, I became an area field representative for the state Republican Party, in charge of running three state assembly campaigns in the California Central Coast Region, headquartered in Santa Barbara. As the political machine revved up for the crucial 1966 elections, Reagan began what we'd call a "listening tour" today, traveling the big state on the banquet circuit as he tested the waters for a run at the governor's mansion. Although I didn't work on the Reagan for Governor campaign, I monitored the race closely.

Back on October 27, 1964, stumping for Goldwater, Reagan had delivered the speech that would later be known as "A Time for Choosing." Entirely self-written, the speech echoed the themes and platforms that Democrats had used when Reagan was a young man. He talked of reducing federal spending, eliminating useless government programs, and returning authority to the states, communities, and people. Now, with his eyes on the statehouse, Reagan took The Speech to any crowd that would have him. It was the beginning of the "common sense" agenda that he would talk about incessantly, and with ever-greater effect, for more than three decades.

The Speech had brought Reagan to the Talk of the Town restaurant in Santa Barbara on that day I first shook his hand. He'd come to meet Republican leaders from the local party for his embryonic gubernatorial race against Pat Brown. Truly a man of the left, Governor Brown was a vocal proponent of all things

government. He was a career politician who looked more like a bean counter than a glad-hander, but he knew how to use the influence of the governor's office to secure his power base. He had trounced Richard Nixon earlier and was salivating at the chance of running against Ronald Reagan. Brown had become California's political Goliath, a two-term governor in a state where, at the time, registered Democrats outnumbered Republicans. Needless to say, the state GOP wanted to take him down.

I was by far the youngest in attendance that evening, and I watched with some interest as Reagan gave his reasons for running and why he needed the help of the local committees. His mannerisms were endearing. He was genuine and a good listener, not typical traits for a politician. He was charming, but it was a boy-ish charm, almost a naïve posture that disarmed you. He clearly didn't have all the answers, but just when you were about to dismiss him, he'd answer a question with such eloquence you were spellbound. He spoke confidently of his "common sense" government, the shortcomings of the Brown administration, and how he envisioned winning the race. Although we met at the end of the program long enough to exchange greetings, I was just another face at another meeting.

Reagan had done all that could be asked of him that evening, but I was still not sold. There was another strong candidate for the GOP nomination, former San Francisco mayor George Christopher, who, as a moderate and proven vote winner, was getting good ink. What's more, he was a known commodity. I'd vote for Christopher if it came down to him against Brown. Winning wasn't everything after Goldwater, but it would be nice not to go down in flames again. As for Reagan, I still felt he was a walking studio set, choreographed down to the last detail, but I knew he had a strong team behind him.

His so-called Kitchen Cabinet was part of the reason Reagan was seeking elective office in the first place. It consisted of Henry Salvatori and Cy Rubel, rock-solid conservatives; Taft

Schreiber, a classic Rockefeller Republican; Holmes Tuttle, an Eisenhower man; and William French Smith, a successful L.A. attorney who was nonetheless the political new kid on the block. These first Reaganites were Big Tent types, representing the diverse factions of the GOP that needed to get along if the Republican Party was going to get back on its feet. The attraction they felt toward Reagan was not really out of any love for the man or his ideology—love and respect would come later. These were good party men first and foremost. They saw Reagan as a team player, a politician who wouldn't alienate any large segment of the GOP and who could rebuild the party. More than a pretty face and a good pol, he was smart, too. And he wasn't a silver-spoon Republican like many in the old Eastern wing of the party. Like his Kitchen Cabinet, Reagan had come up from the trenches. Holmes Tuttle had ridden a boxcar to California in the 1930s. Before he became the largest Ford dealer in the world, he used to sleep in the cars he sold to make ends meet. Justin Dart, another of the early Reagan advocates, was self-made, too, a drugstore entrepreneur who had expanded his interests globally. I think these party wise men at first thought they could control Reagan, but they would ultimately discover that he followed his own drummer.

I yearned for the same thing as the Kitchen Cabinet: I wanted to win, too. As Reagan worked the state, I was hard at work recruiting candidates, managing press relations, raising money, and developing creative advertising concepts. The conventional wisdom within the party was that Reagan had a chance to unseat Brown and that his coattails might be long enough to drag a lot of people to Sacramento with him. Not only was his message getting through to the people, but Californians were going through a sort of "Brown fatigue." He'd been there too long and every common campaign trail gaffe became magnified.

During the campaign, Reagan made it clear that he was a party builder. He worked on behalf of candidates statewide and opened both his wallet and his organization. Instead of acting like an actor, Reagan was showing the instincts of a seasoned politician. All of a sudden, he started to make sense to me.

Democrats, though, weren't about to roll over. Throughout the state they were following Brown's lead, relishing the chance to go against Reagan in the general election. It was their first miscalculation about the effectiveness of Reagan—refusing to accept the fact that this ex-actor could actually govern, labeling him a lightweight, claiming he didn't have the requisite experience to manage a bureaucracy. And it was a miscalculation that would be repeated again and again. One of Reagan's greatest advantages in his political career was that he was forever underestimated.

Meanwhile, the three assembly races I managed looked pretty good, and I was beginning to feel pretty cocky when one of my best candidates bailed out on me. It was the spring of '66; the filing deadline for candidates was only weeks away. I was eating a late breakfast in Santa Cruz's Bubble Bakery, a café I frequented, while the short-order cook, an armchair political strategist, commiserated with me over a cup of bad coffee.

We had a fairly weak Democrat in this district, I told him. If I just could field a body, any body, I might have a chance. I literally picked a man out of the restaurant that looked like a good candidate. He was chatting away in a booth with a few people. I think I had seen him around, but I didn't know his name or what he did for a living. Being a visuals guy, I just knew that he'd be good on television. After all, Jack Kennedy taught us that being telegenic was half the battle, and this guy definitely looked Kennedyesque.

I ran my idea of desperation quickly past my friend the cook, and he just as quickly burst my bubble. The kid I'd settled on—his name was Frank Murphy Jr.—was the son of one of the area's biggest Democratic bosses. Undeterred, I gave Murphy my card on the way out, saying that if he is interested in running for GOP office, he should give me a call.

He called an hour later. Without missing a beat, he said he'd do it but couldn't move forward without Frank Murphy Sr.'s okay. At the time, his father headed up the county Democratic Party, and the two of them worked together at a small law firm in Santa Cruz where, together, we pleaded our case to the old man. He listened attentively to Junior, then me. I told him that I would be lining his son up with some of the best political support in the region and that he would be getting solid funding and advertising support from the state party. I think that Senior may have known that his son had political aspirations, but it became clear that they had never talked it over. Now, his kid comes in with a field hack from the Republican Party and is all of a sudden ready to jump in feet first. I think it threw the old man for a loop. He didn't say a word, though. After I finished, he stood up and shook my hand, while looking at his son. He said, "Frank, if you're ever going to run as a Republican, this is the time to do it. This is their year and it's all because of Reagan—he's going to deliver."

That November, I went on to win two out of three of my races and, just as Mr. Murphy had predicted, Ronald Reagan carried the day. With him he took young Murphy, who served nine years in the state assembly. Reagan not only ate Pat Brown's lunch, he did so decisively, beating him by more than a million votes.

Back in my little world, I had become fascinated with the power of advertising and direct mail in political races. I had relied on creative ads extensively in my three campaigns and my immediate plans were to hunker down in Santa Barbara, joining a small advertising firm. Unlike most field men, I had little interest in joining the governor-elect's team in the capital city. I loved advertising. I loved living in Santa Barbara. I simply could not imagine living in Sacramento, a sweltering valley town.

A few weeks later, my plans for becoming a creative genius in the ad world were put on hold. The call that would change my plans—and my life—came from Denny Carpenter, Reagan's new handpicked vice chairman of the state GOP. He directed me to report to William P. Clark, the head of Reagan's transition team. I don't know why Denny made the call to me. If he was told to do it, it was (and remains) a mystery to me who would have wanted me on board. But there I was, Clark's number two on the transition team. I came to genuinely like and respect Bill Clark, and I served him well. After Reagan took the oath of office and appointed him secretary of the cabinet, Clark told me that I'd need to move up to Sacramento permanently. He wanted me to be his assistant cabinet secretary.

I accepted the offer. By now, I was enthusiastic about Reagan's programs although I still found it hard to believe that he was actually the governor. I understood that he campaigned on the notion that he'd be a "citizen-governor," but I wondered to myself how effective a man with no elective experience would be in handling one of the nation's toughest political jobs.

Six months into the Reagan administration, there was a shake-up in the inner circle, and Reagan appointed Bill Clark as his new chief of staff. Bill made me his assistant, and I found myself as the figurative, if not the actual, number two man in the office.

I had spoken no more than five words to Governor Reagan since seeing him two years ago at the Ambassador Hotel. He still had no idea who I was despite the fact that I'd be moving just a few doors down from his office. I'd said even fewer words to his wife, but I was told that I would be her contact within the inner circle.

Bill plunged into the complex areas: the budget, appointments, and all the serious responsibilities that come with governing the Golden State. My official job description called for me to oversee the governor's schedule, political liaison, events, and special projects. I soon learned that all of these areas were of special interest to Nancy Reagan. This has remained the case through Sacramento, Washington, and today.

From the inside, the reviews on Nancy were not pleasant. Many who had dealt with her said she was at best demanding, a tough-minded political wife who needed constant attention. Not surprisingly, then, the first phone call from Nancy Reagan was the most terrifying experience of my nascent professional life. The call was patched through from her home in East Sacramento, and within seconds the receiver was laced with sweat from my clammy hands. Thankfully, we began by exchanging minor pleasantries. Her voice was far from intimidating, but she quickly ended the small talk when I mentioned the weather for the second time. In a no-nonsense manner, she got down to business, asking me to change the governor's schedule to accommodate an upcoming trip she was planning to Los Angeles.

She wanted "Ronnie"—it was the first time I'd heard him called that—to make the trip with her for reasons she didn't disclose. I managed to advise her that the assembly was in session that day and it would be a terrible idea for him to be out of the capital. If she really needed him to go with her, she really should consider rescheduling her trip for when the assembly was out of session.

There was dead silence on the other end of the phone—a five-second pause that felt like a half hour before it ended with an understated,

"You're probably right." We exchanged a few more words, and I hung up. Relieved and surprised, I reflected on how it had gone better than I'd expected. Maybe I'd have no use for the supply of Pepto-Bismol stashed under my desk, earmarked for my dealings with the first lady. After all, she had been entirely rational and accommodating. Then it dawned on me: the first lady and I had the same interests— helping her husband do the best he could and to look good while doing it. I learned that the secret to working with Nancy Reagan was to employ a radical tactic: treat her like a human being. Most important, tell the truth. This has been the foundation of our relationship, and from that day on, I never once worried about talking to Nancy, either in person or on the phone. I began to see her as another member of the staff, a person who would have good ideas and bad ones, but who was always thinking about the endgame.

Of course, I wouldn't gain her trust in one phone call. She was very protective of her husband and wanted to make sure that I had his best interests in mind. I wasn't out to make headlines; my name would never be in the paper. (How naïve I was!) I was happy working for Ronald Reagan and didn't need the recognition. Once she realized this—and it didn't take long—we formed a bond that has lasted ever since.

My relationship with Nancy made me better prepared both to serve Reagan and to develop my own political skills. By taking the job that nobody else wanted—the aide who, in addition to other duties, would have to answer to California's inexhaustible first lady—I had found my niche, and a lifelong friend.

The insight I gained from Nancy early on proved helpful in ways I can never count. It was almost like a catechism: I'd tell her what I was intending to do to edge the governor in this direction or that or to achieve some political end, and she would patiently correct me— "No, Mike, that won't work"—and then explain exactly why. I learned to listen carefully because it quickly became apparent that absolutely no one knew Ronald Reagan better.

Nancy taught me ways to win over the governor, ways that other aides were unaware of. If I was to ever prevail upon Reagan, she would advise, I could never use blatant politics as a tool to pull the governor in my direction. If I said that going to a certain event or supporting a certain bill would mean "political death" for him, he would dismiss my argument out of hand. But if I said that his support of this bill or his attendance at this event would hurt people, Reagan would demand to know more and usually take my side if I could prove my case. Few of his closest aides realized this, but his most important ally, Nancy, knew it was often the only way to move an inherently stubborn man.

Reagan never once questioned my relationship with Nancy. As the months and years went by, I think he may have been comforted by the fact that his wife had a confidant within the inner circle—one who liked and respected her.

I made sure that Nancy Reynolds, the first lady's press secretary and friend, had an office next to mine, a symbolic gesture that elevated the importance of the first lady. Eventually,

Reynolds and I would accompany the Reagans nearly everywhere they went.

Even with Nancy's counsel, though, most of my most important early lessons in dealing with Reagan came from the man himself. Californians spent the first months of 1967 warming up to their new governor. He was implementing his agenda and was high in the polls, still enjoying his honeymoon with voters. Although I remained "what's his name" in the governor's lexicon, I was starting to grow into my position.

Bill Clark told me that I'd need to get smart on a couple of issues because I'd be briefing the governor Monday morning for a speech he was giving at a chamber of commerce luncheon. I was anxious but excited that I would finally get the chance to prove my worth.

I had heard that Reagan was a softy on his staffers and that he'd probably give me one of his patented nods and offer a rhetorical pat on the head. Still, I wanted to make my mark as a valued part of his team. I spent the weekend plotting my approach to my first gubernatorial one-on-one.

As a hardened veteran of chamber of commerce speeches, I knew how these lunches worked. Between forkfuls, attendees got to write questions for the speaker on three-by-five-inch cards. They would then pass them to the chamber host who would read five or six of the questions to the guest during a question-and-answer session. I took the liberty of rigging the process. Cashing some chits in with friends, I arranged to provide the chamber with the questions the governor would be ready to answer, assuring everyone that he'd use the opportunity to promote the chamber's short-term goals.

On Monday morning, I strode into Reagan's office proud as a peacock and coolly outlined what I had done over the weekend. I handed the governor the six cards containing the questions. He shuffled through the cards without looking at them. The smile on his face turned into a semifrown, a look of fatherly disappointment. It was the same look I used to get from Mom after lifting a five spot from my dad's wallet.

He looked at me and said, "Mike, this won't work."

My heart sank. I had laid an egg on my first assignment. Instinctively, noting my despair, the governor smiled and said, "You can't hit a home run on a soft ball." The cards found a home on the bottom of his trash can.

The workshop was just beginning, however. I had a lot to learn—and Reagan had a lot of teaching to do. It was barely three days later when word came down that a rising-star newsman was planning on filming a "Day in the Life" segment with the new governor. The reporter, a guy named Tom Brokaw who worked at Southern California's KNBC, had told us that people were hungry to see the "real" Reagan.

We knew the press would continue to hit the "right-wing actor turned governor" angle as long as they could get away with it. We were taking a chance with Brokaw, but our press people said that he was one of the more objective journalists around, so with the governor's approval, we agreed to do the segment. Feeling that I needed to dig myself out of the shallow grave I had dug for myself on the chamber of commerce incident, I persuaded Bill Clark to let me brief the governor on the Brokaw piece. He said the project was all mine. He had obviously heard not a peep from the governor about my first briefing.

Again determined to make a splash with Reagan, I spent another weekend planning a comprehensive strategy. My second attempt at genius would not fail me. Knowing that he was an ex-actor, I thought he'd appreciate a well-scripted event, so I choreographed the day down to the last minute.

At 9:03 the governor would walk down the steps into the Capitol Park; at 9:07, he'd take his jacket off and sling it over his right shoulder, just the way Jack Kennedy showed us all how to do it. The script went on and on until 4:05 P.M., when I would have the governor shaking the young anchor's hand.

It was Monday morning in the governor's office, and Reagan earnestly reviewed the six-plus feet of butcher paper that lined his walls. It was my comprehensive outline of the entire day, complete with maps and mock drawings of the relevant characters. At first, it appeared that he was getting into things. He nodded when he got to the 9:03 entry—the first one—but cringed when he saw the mock-up of him slinging his jacket. Finally, he stroked his chin and rubbed the back of his head before saying, "Mike, I really appreciate your trying to tie all this stuff down, but I can't do this." He took my baffled expression as license to proceed.

"You're asking me to do things that I'm not comfortable doing. If I'm not comfortable, the people at home watching this aren't going to be comfortable, either."

Professor Reagan had just given me my second lesson in as many weeks. Nine out of ten politicians would have been thrilled with this type of advance work, but not Ronald Reagan. He had spent a lot more time in front of the camera than most and implicitly understood what worked.

I was grateful that he never talked down to me in those early years. I guess he would have lost patience if I had repeated my mistakes. But I never did. Everything I was to do for Reagan in Sacramento and later in the White House would come from a burning desire never to disappoint the man, no matter how insignificant the job at hand. As a boss, there was just something about him that made you want to please him and do your best. This applied to everybody—everybody—who worked for him, regardless of age or position. Reagan's housekeeper in Sacramento, a big German woman named Ann, would regularly fill his briefcase with treats whenever we went on the road in those early days. Cookies, jelly beans— she had his sweet tooth down pat. But the special attention didn't come only from those who were paid to make him comfortable. All of us on his professional staff—in Sacramento and later in Washington—responded to him almost as if we were seeking a father's respect and judgment. Even newcomers would feel the pull. Richard Darman, Reagan's brilliant and valuable White House aide, hardly seemed the type to melt in the president's presence, but he did time and again. Reagan rarely tossed out bouquets, but you always knew where you stood with him, and he knew how to get the most out of people.

Martin Anderson, the economist, recently told me a classic example of how Reagan would subtly get his message across. It was the mid-1980s and Secretary of State George Shultz was just wrapping up a meeting with the president. Before leaving, the scholarly Shultz asked President Reagan if he had time to take a look at a speech Shultz had prepared for an address he was giving over the weekend. Reagan said he'd be pleased to look it over.

Later in the week, Shultz was wrapping up another meeting with the president when he remembered his speech. "By the way, Mr. President, did you have a chance to review that speech?"

"Oh, yes, George, I looked it over. Not a bad speech," Reagan said with a smile. "But I wouldn't give it." It might need a little more work, George, that infectious grin was saying. Point made: the two of them sat down and rewrote it together.

Reagan took a lot of heat from some quarters for being a hands-off manager. When he was first elected, critics went out of their way to note that he had little management experience and none in running a major bureaucracy. Reagan heard these criticisms, but he didn't buy them. While he didn't hold an extensive management portfolio, he took great satisfaction in having served six terms as president of the Screen Actors Guild. Furthermore, he had spent considerable time learning best practices from senior managers at General Electric, the place he called home soon after his acting days. And his passion for reading helped him glean insight from biographies of successful managers and politicians.

Reagan's management style was simple, and

it never changed during all his years in office. It was successful for three reasons. First, he both attracted and demanded the best people. This may sound like common sense, but in politics sometimes the best and brightest are on the other side or have other disqualifying issues. In 1980, when it came time for him to consider the most important staffing decision of his new presidency, the chief of staff appointment, many folks considered longtime Reagan policy hand Ed Meese III and possibly even me to be the frontrunners.

I recall discussing the appointment with Reagan a few days before the general election at the candidate's rented farmhouse campaign headquarters in suburban Virginia. Over an after-dinner cocktail, he brought up the subject of potential candidates for national security adviser.

"Governor," I said, "before you think about that, you need to think about a chief of staff."

"Well," he said, nodding in agreement, "what about Ed?"

"Ed is great," I said. "But I think you should

also consider someone who knows Washington, knows the way the town works, and Ed has always been most valuable as a counselor. I know you rely on him completely for advice, and you'd lose that if you asked him to run the White House day-to-day."

He asked for my ideal candidate. I suggested James A. Baker III, the man who ran George Bush's campaign against us in the primaries. It says a lot about Ronald Reagan that he didn't laugh me out of the room. Here I was suggesting a guy for the most important job in the administration who not only worked against us in 1980 but also had been part of the Ford operation in 1976. If there was any Republican who didn't seem to be a Reaganite, at least back then, it was probably Jimmy Baker.

Instead of laughing, Reagan looked at me and asked, "Do you think he'll do it?" If he was going to go outside the circle and get somebody, he may as well get the best. In Baker, he got just that. I'm not sure Ed Meese ever forgave me for the recommendation, but the combination served the president well, and Ed and

Jim both moved on to cabinet posts in Reagan's second term.

(I've told this story before, but Jim Baker also inadvertently caused me one of my most embarrassing moments. Back in 1976, a Ford operative complained on television that Reagan wasn't doing enough to support Jerry Ford's reelection. When Reagan heard the charge, he told me, "Get Baker on the phone!" I did, and Reagan lit into him, telling him how hard he had been working on the president's behalf. One problem, though: I'd rung up Tennessee senator Howard Baker, not Jim Baker, and Howard hardly knew what to say.)

The second reason Reagan was a good manager is that nobody threatened him. Nobody. Some people, regardless of position, are fearful that a subordinate might outshine them. Reagan was just the opposite. He hoped that his subordinates would be smarter than him. If he could bring in a young genius with a more muscular IQ, so be it.

The no-threats rule extended to the opposition as well. Shortly after the president moved into the Oval Office, he was rummaging through its famous kneehole desk and came upon a leftover from the Carter administration: a medal that had been bestowed on Robert Kennedy but never delivered by Jimmy Carter, who didn't believe in presenting such awards. The easy thing would have been to have dispatched an aide over to Capitol Hill to drop the medal by the Senate office of Ted Kennedy. Instead, the president got Ted on the phone and asked if he thought a delegation of Kennedys would like to come by the White House for a small ceremony. A few weeks later, the Oval Office was filled with a high-spirited Kennedy clan, seemingly delighted to be back in the building that had made their family name anathema to the conservatives who formed Ronald Reagan's core constituency. The seemingly strange relationship didn't end there, by the way. When Ted Kennedy needed a star to help raise funds for his brother Jack's library at Harvard University, he asked the president if he would lend his name. Of course, came the answer, just as I knew it would.

❦

Finally, Reagan firmly believed that it was his job to set the priorities of his administrations and to make the big decisions. This would be delegated to no one, and it was the same in Sacramento as it was in the White House. He took these responsibilities very seriously. Because making the big calls weighed heavily upon him, he developed an ironclad system that proved extremely effective in helping him deliberate.

If a major new issue demanded Governor/President Reagan's attention, a few things would have to happen before he'd settle on a position. Reagan would demand to see everything that had been written on the issue. Aides invariably would include the requisite executive summary on top of the pile of information, hoping that Reagan would at a minimum digest that overview. Much to their chagrin, though, he would read everything they gave him.

Once they learned that he read everything, these same aides naturally came to fear leaving anything out, and the piles soon doubled in size. Still Reagan would digest all the material. At the same time, unbeknownst to us, he'd also be picking up tidbits of information on the issue through other channels. Sometimes he'd cut out an editorial from a small weekly newspaper or glean information from a constituent's letter or courtesy call. These things all went into the Reagan memory bank and stayed there.

Early on in the White House, I used to caution colleagues who didn't know Reagan well, especially the erudite Jim Baker and Richard Darmen, that they needed to be extremely careful about the information they passed on to Reagan. Anything they gave him would be entered into his mental computer and could be spit back out at any time in the future. Reagan remembered nearly everything he read and most of what he was briefed on.

After he'd gotten all the information and digested it, we'd assemble the cabinet or the top staff to discuss the pros and cons. Never once did Reagan ask, "What do you think I should do?" He would pull information from each of us in the form of questions, many of which came from his stack of briefing papers but also from unknown sources. He would next have a smaller session with his closest aides. In the White House, that meant me, Baker, and Meese. He'd let us know the way he was leaning, and then he or someone else would make his announcement at the next cabinet or senior staff session. Once he made his decision and it was implemented, he never once looked back. He believed deeply that this process worked, and I never heard him say he regretted one of his major decisions.

The beauty of this system was that it moved at Reagan's speed. Depending upon its urgency, this process could be played out in a matter of days or just in a few hours.

Because of my previous work in the field for the Republican Party, I assumed the role of principal political coordinator in the governor's office. Neither Clark nor Meese had political backgrounds and both felt more comfortable handling the day-to-day business of the office.

That meant that dealing with party officials, coalitions, and volunteer groups fell to me. The job could have been a gulag, but fortunately for me, political outreach was key to Reagan's life outside of the state capitol. Ronald Reagan was a hot ticket not only within California political circles but was sought by countless conservative and Republican groups all across the land. It was my great fortune to travel with him to nearly every one of these events—hours spent talking, listening, and learning from Ronald Reagan up close and uninterrupted. With each trip, I became closer and closer to the Reagans, until I came to feel a part of their extended family.

It was during this time that I met my wife and life partner, Carolyn Judy. She came to work in the capitol in 1968 after serving in the governor's San Francisco field office. The first time I saw her, I knew she was the one, and I've never had a moment of regret. We were married at Yosemite Valley in July of the same year and bought a home a few blocks away from the governor's residence in East Sacramento and began a family. Our two children, Blair and Amanda, were born there and life was good. These were wonderful times for me. Not only did I have a growing family, but also a front-row seat watching the political development of a soon-to-be legend.

In Nancy, Carolyn also found a good friend. They had known people in common before we were married, and they shared a similar sense of fashion and taste. In fact, some people would remark that they looked alike. (Years later, in Washington, Carolyn would be pumping her own gas one day when a man pulled up beside her and gushed, "Gosh, Nancy Reagan filling her own tank!") In time, Carolyn would join Nancy Reynolds as the First Lady's traveling buddy and weekend companion at Camp David. On both fronts, the two of us felt deep ties to the Reagans.

I'll never forget the day I told Ronald Reagan that Carolyn was going to have a baby. "Pray for a girl," he said.

"What about Ron?" I asked, referring to his son, who was still living at home at the time.

"Oh, I love Ron," he said, "but having a little girl is like seeing your wife grow up all over again."

Three decades later and counting, he still couldn't have been more right.

CHAPTER TWO

THE CAMPAIGNER

To learn what kind of campaigner Ronald Reagan was, look no farther than his long-shot race for president in 1976. His primary challenge against President Gerald Ford reflected not only his never-say-die optimism, but also the fierce competitive streak that lurked behind that gentle smile.

The race was an emotional ride for Reagan and the small cadre of aides who accompanied him on that quixotic mission. We had gone toe to toe with Ford throughout the primaries—losing New Hampshire by just over a thousand votes. Could Reagan really do it? At the time anything seemed possible, including beating a sitting president.

He didn't, of course. Our Waterloo came in Florida, where Ford won and seized the momentum. He would win the next several primaries and drive a stake through the heart of the Reagan campaign, but not, we would soon find out, through Reagan himself. The campaign was a million dollars in the hole, and the senior staff was worried that Reagan risked embarrassing himself by staying in the race. Even Nancy Reagan urged her husband to step out gracefully.

Reagan would have none of it, saying he would see Ford at the convention. He must have known something that the rest of us didn't. We limped into North Carolina with no hope, using money we didn't have to buy a half hour of airtime for Reagan to tell the Tar Heel State why he was still a viable candidate. On the surface, it looked nuts. Deeper down, it looked even nuttier. Worse, the first taping for the spot we couldn't afford was absolutely awful. The Great Communicator looked fatigued.

The day-by-day, night-by-night grind of the campaign was taking its toll.

After the taping, I overheard the director giving Reagan a rave review of his performance. I knew better and headed back to the studio's kitchen to make up a pot of real Navy coffee, double the caffeine. Reagan rarely drinks regular coffee, opting instead for those little bags of Sanka decaffeinated you see at bad motels. When the Navy coffee was brewed, I cut in on the director, offered my unvarnished opinion of the speech, and handed Reagan the double-barreled cup of coffee. I had used this trick a few times in the past when I noticed Reagan off his game. It was rare to see Reagan deliver a lackluster performance in front of the camera, but even the Gipper couldn't turn it on every time. During the first take, he knew in his heart that he didn't deliver the goods and understood my thinking when I handed him the java. No words needed to be exchanged. After he drank it down, we agreed on a retake.

As he sat before the cameras a second time,

he was the old Reagan, animated and lively. His head bobbed and his cheeks were rosy. Of course the coffee was just a reminder that he could and would have to do better. It was just too important and he'd need to make Take Two count. I think he gave the best version of The Speech that I heard during the entire '76 primary season.

The Republican voters of North Carolina agreed, and handed him their prize. It may have been the most important political victory of Ronald Reagan's life. The Big Mo was back. Reagan would go on to win every precinct in then Democratically controlled Texas, changing the politics of that state for generations to come. More victories followed: Alabama, Georgia, Indiana, Nebraska, and, of course, California along with several other western states. The primaries would end with Ford only 136 delegates ahead of Reagan.

We weren't out, but we were still down. It was time to try to capture some attention from the national media in hopes that the delegates in play would keep us in mind. Our bold stroke, proposed and pushed through by campaign manager John Sears, was to announce our running mate a month before the Republican convention in Kansas City. Pennsylvania senator Richard S. Schweiker, a decidedly moderate-to-liberal politician, was Reagan's choice as vice president. Plenty of differences on issues separated the two men. Schweiker himself was quick to point out many of them when he and Reagan met for the first time to discuss the possibility of Dick's being on the ticket. As always, Reagan respected honest differences openly stated. More important, I think, he simply liked and admired Dick and his wife, Clair. They had grown up in a small Amish sect in Pennsylvania, and while both had moved out into the wider world in the years since, they had retained the quiet strength of character taught by their singular upbringings. And, of course, we hadn't just picked Dick's name out of a hat.

Our intent was to start a domino effect, using Schweiker to help persuade the officially "uncommitted" delegates of Pennsylvania to join

Team Reagan. Ideally, after the dam broke in Pennsylvania, neighboring New Jersey would fall into our column. Our hope was to have Pennsylvania GOP powerbroker—and committed Ford backer—Drew Lewis come our way. But Lewis stayed committed to Ford, and our strategy foundered. Tellingly, Reagan picked Lewis four years later to serve as his secretary of transportation. The man is incapable of grudges.

Burned in the mid-Atlantic states, Reagan and Schweiker then looked to Reagan's own wavering delegates in the South. But from Richmond to Atlanta, Birmingham, and beyond, the liberal Schweiker was the Antichrist. The conservative base was outraged with Reagan's choice of running mate. They held their fire initially, but soon after they piled on with an undeniable zeal. I took a call at 4:00 A.M. from one of our true believers who asked, "Have you counted your thirty pieces of silver yet, Judas?"

Most conservatives still loved the Gipper and wouldn't say anything as rude to his face, but there were other ways to get the message across. In one particularly memorable case, we had gathered a small group of wavering delegates in a hotel conference room in Mississippi. Reagan was seated at the dais with Senator and Mrs. Schweiker, both devout Christians. I'll never forget the following exchange:

"Mr. Governor, may I speak?" a delegate asked. He was a sophisticated-looking fellow, decked head to toe in traditional Brooks Brothers garb, portly with a round, bespectacled face. The words he uttered betrayed his dapper appearance.

"Of course," Reagan said.

"Ah don't know how you could have this fella," the delegate started. "Ah'm not a drinkin' man, but the night ah heard you picked Dick Schweiker ah went home and drank a whole pitchah of whiskey sours. I would rather my doctor had told me my wife had a dose of the clap."

The Schweikers looked to the floor, blushing. Reagan, unfazed, pretended the delegate had asked a completely different question and casually steered the discussion into the many things that he and the senator agreed upon. Then he coolly outlined his strategy for getting to the convention and told how proud he was to have Dick on the ticket.

It was like this for weeks, during which time Reagan exhibited remarkable strength and perseverance. He was always the first man on the campaign plane in the morning and the last guy to hit the rack at night. (I wish all the people who wrote that Reagan was lazy could have spent that spring and summer with him.) Undaunted, Reagan believed that he had to compromise on some things to keep things moving forward. Schweiker's liberal bent mattered little to Reagan. He knew that getting to the convention was the next step in reaching the ultimate prize. If he had to take a little heat, it was worth it. His followers just needed to trust him.

And, in fact, Reagan was right. Thanks to Schweiker and the buzz he generated, the media believed Reagan was still a viable candidate when we arrived in Kansas City for the convention. Enough delegates were undecided and Reagan was hustling for their support. If he had not bet on the Pennsylvania senator, the press wouldn't have had a reason to talk about Reagan for weeks, and that would have all but killed the campaign. In presidential politics, if the press doesn't think you're alive, you're not. It's as simple as that. We made it to the convention but in the end lost to Ford and the power of the presidency. Ford had leveraged himself and his office, pulling the uncommitted delegates to his side of the aisle with promises of roads, jobs, and appointments that only a sitting president can make. Reagan's people were livid, but Reagan wasn't. His inner calm amazed even me. He showed magnanimity toward Ford even as he filled the role as comforter in chief for his own morose campaign staff.

Nor did Ronald Reagan ever forget that when things started falling apart at the Kansas City convention, with conservative delegates jumping ship left and right, Dick Schweiker had come to him and offered to step down so he could pick a new, last-minute running mate more to the right's liking. Reagan wouldn't let him, and four years later, when it came time to name a secretary of health and human services, he picked Dick over a half dozen more qualified or logical contenders. After all, the two had been tested by fire together.

The night President Ford was to give his acceptance speech, Reagan sat with Nancy in a box overlooking the confetti-filled stage. I was standing in the back of the box near the door, watching CBS's Mike Wallace interview the governor, when John Sears escorted in prominent presidential adviser Bryce Harlow.

"Jerry wants Ron to join him on the stage," Harlow proclaimed.

Reagan had repeatedly told anybody who would listen, "This is Jerry's night, and I'm staying in the box. He's earned it." Despite this, Ford's man insisted. I repeated Reagan's position and reminded them that the national media would eat us alive if Reagan came down. They'd write that he couldn't accept defeat and was raining on Ford's parade. "The only way the governor's going down there is if Ford asks him in front of the entire world," I said. With Ford making the overture on national television, Reagan would be inoculated from charges that he was trying to steal the spotlight.

About five minutes later, Ford looked up to the box and beckoned his longtime rival down to the stage with a "Come on down, Ron," that left Reagan no excuse. Reagan pulled me aside briefly as he walked out the door. "Mike, I have no idea what I'm going to say."

"I'm sure you'll think of something," I said, confident that he would rise to the occasion in front of the biggest audience of his life at that point. And did he ever.

Reagan proceeded to give a roof-raising speech that left countless delegates with feelings of regret. Given without preparation or notes, the speech was one of the most eloquent I had ever heard Reagan give. It was filled with lines that can still inspire tears and goose bumps—a collection of his best stuff, including some I'd never heard before. He didn't talk

about himself or how he lost the campaign—
that wouldn't have been Reagan. I don't think
he even talked about Ford. He was driven by
something much bigger than any one man, and
he said so. "We live in a world in which the
great powers have poised and aimed at each
other horrible missiles of destruction that can,
in a matter of minutes, arrive in each other's
country and destroy virtually the civilized
world we live in."

Later that night most of Reagan's senior
aides sought to kill the pain, using a strong pre-
scription from the governor's minibar. We
griped and talked into the night about what
next and new jobs, while the Reagans turned in
about 12:30, slipping into the bedroom of the
huge suite we were using for convention headquarters. My wife, Carolyn, and I eventually
went to sleep in a smaller adjoining room.

The next morning, I was startled out of bed
by pounding on our door. When I cracked the
door, I found a few of Reagan's patrons from
the Kitchen Cabinet: Justin Dart, Holmes Tuttle, and Bill Smith. They were all wearing iden-

tical outfits—gray flannel pants, blue blazers, rep ties.

"Where's Ron?" demanded Dart.

"He's asleep," I said. "It's seven-thirty A.M."

"Get him up," one of them barked.

Realizing my confusion, Holmes Tuttle said, "Look, Mike, we've been talking with Ford's guys all night. We want to talk to Ron about the vice presidency."

I crept into the governor's bedroom, where he and Nancy were in deep sleep and tried to make some noise to wake them.

"Holmes, Jus, and Bill are here and want to see you," I said. "They want you to be Ford's vice president."

He rolled back onto his stomach, speaking into his pillow. "Tell them I don't want to."

Knowing that they'd never take my word for it, I told him that they were his friends and that he'd have to tell them directly. Reagan got out of bed and quickly dressed, and I moved to the kitchenette to make some coffee. In very little time, Reagan bounced into the room. As he greeted his visitors, the phone rang. It was the

White House switchboard advising that President Ford was calling for Governor Reagan.

I motioned to Reagan, and he picked up another phone. There was a series of brief pleasantries, about a dozen "yesses," and half as many "fines." The call ended with Reagan saying, "That's terrific. I think you made a wonderful choice."

Reagan hung up and approached his friends who were waiting anxiously. "That was Ford. He just told me that he's picked Bob Dole as his running mate."

The three men who had been Reagan's most ardent supporters and patrons looked at each other. They were devastated. I realized then that these tough, successful businessmen were softies after all. Their eyes welled and tears fell. Their plans for saving the Republican Party appeared dead with the demise of the "Ford-Reagan" dream ticket. Deep down they surely thought Ronald Reagan was done forever.

Reagan, used to his role as comforter by now, hugged each one of them at length, patting them softly on their backs. His shoulder was wet with the tears of three mature men. Smith and Tuttle left first, leaving Dart alone with Reagan and me. As Reagan was slowly showing his last visitor the door, Dart opened his mouth to say something but was cut off by Reagan's raised index finger.

"Remember, Justin, only the lead dog has a fresh view. We'll have our day."

Dart and company missed a small talk Reagan made to his friends in the California delegation later that morning. Amid the sniffling of his admirers and staff, he remained stoic, paraphrasing an old saying: "I may be wounded, but not slain. I'll lay me down and rest a while, and then we'll fight again." I had been standing in front of swinging doors at the side of the room. When he said that, I slipped through the doors and finally wept myself, alone in an empty hallway.

These are not the words of a man pining for the vice presidency. Ronald Reagan would not spend four years looking at Gerald Ford's posterior. He had things to do and places to go. The 1976 campaign would soon be a distant memory. He would be back.

Unlike the most recent crop of presidential contenders, Ronald Reagan hadn't grown up partly on the campaign trail. Al Gore and George W. Bush didn't have to be told what hard work campaigning is; they just had to look at their fathers. Reagan had all the tools to become great at it, but he first had to master the immense preparation and discipline that campaigning requires. And he had to figure out exactly what political party he belonged to.

As political junkies know, Ronald Reagan was a liberal Democrat during his young adulthood and early middle age. Refusing to abandon the working-class roots of his Irish father, young Reagan often described himself as a "Roosevelt man." Franklin, not Teddy. Reagan always told me that if I wanted to truly understand what he believed back then, I should go read the text of FDR's first inaugural. He'd grown up in a household where FDR was a figure worthy of worship, and Roosevelt would continue to serve as a role model for Reagan throughout his political life.

Skeptics will point to the rich irony. Reagan came to the White House vowing to slay the federal leviathan, a beast launched in large part by his idol, FDR. But whatever he later came to stand for, Reagan was a product of the Great Depression; he saw his dad lose his job as a shoe salesman only to be saved by FDR's WPA. Reagan would never come out and openly reject FDR's policies; he would simply venture that his America had—or at least should have—grown out of them. In his early days, he was a true party man, but eventually switched to the GOP. While many critics blamed Reagan's switch on his admiration for Jack Warner, other Hollywood corporate types, and his father-in-law, surgeon Loyal Davis, I've always believed him when he said of the Democrats, "The party left me."

If Reagan's transition from Old School Democrat to Mr. Republican was evolutionary, the moment that marked his arrival as a GOP torchbearer—the 1964 speech in defense of Barry Goldwater and conservatism—was decidedly revolutionary. To the men who would later comprise the Reagan Kitchen Cabinet, The Speech was the communiqué of a rising force. With his talk of a "rendezvous with destiny" and his description of the United States as "the last best hope of man on earth," Ronald Reagan sounded like a man who could inspire and lead a party in disarray. The rub was that the man who uttered those words was not completely aware of the raw emotions he was touching within Republican circles. Although he considered requests to run for Congress twice, Reagan knew little about the machinations of state government and never really thought about running the state, let alone the nation.

If any single entity was responsible for pushing Reagan toward elective office and his new philosophy, it was General Electric, which hired him as spokesman and had him speak to GE employees at dozens of facilities around the nation. Reagan also hosted a television show for the company, *General Electric Theater*, which ruled Sunday night ratings like *60 Minutes* does today. It was "appointment television." It was while he worked for GE, from 1954 to 1962, that Reagan began to develop a particular distaste for spiraling income taxes and big government, a pair of issues that would endear him to the conservative movement.

GE was also the perfect training ground for Reagan the politician. As the company's corporate face, he was paid to meet with GE employees across the nation, many of them blue-collar workers whom the GE leadership wanted to instill with the virtues of the free market. The brass was convinced Reagan's star power would bring the men and women of GE out to hear a message of why less government intervention in the marketplace was better for business. At receptions and plant tours from coast to coast, Reagan delivered the message, but he also learned how to listen, how to speak his mind without offending, and how to pace himself on the trail—priceless lessons for a would-be politician.

Although it would be a number of years until Reagan gave The Speech, he essentially said much of the same thing time and time again during his ride with GE. To be sure, his words were not controversial or overly political, but they were precise, common-sense iterations on the inefficiencies of central planning and the virtues of liberty and freedom. For the first time, the man who had stood before movie cameras now had to respond to a living, breathing audience. It was a two-way street—not simply a recitation of lines he had locked into his memory.

Reagan initiated a robust question-and-answer session in all his employee addresses, and thus he also learned how to master the art of connecting with his audience. Reagan told me that he would walk away from each session a smarter man, adjusting his speech accordingly. He became well versed in what stoked the passion of a wide cut of American voters, particularly working ones. The rough-and-tumble give-and-take of his question-and-answer session became the ultimate focus group. High-priced polling research couldn't compare to the feedback he heard directly in the field. Later, many critics would marvel in disbelief at Reagan's ability to connect with the average American voter. The truth is, he was an average American, and he would spend years listening to what made them tick. And a lot of the credit for that goes to General Electric.

Reagan, after all, had come of age in Hollywood. Whether he wanted to be or not, he was part of the American entertainment elite. Bill Holden and his wife had been best man and matron of honor at the Reagans' wedding. Their movie friends included the likes of Jimmy Stewart, Jack and Mary Benny, Greer Garson, Efrem Zimbalist, and Robert and Ursula Taylor. I don't think I've ever seen Reagan more broken up than when he gave the eulogy for Taylor, who had replaced him as host of *Death Valley Days*. It was rare air he had learned to breathe. Had he never been forced to reach out to the employees of GE, he might well have foundered when he finally got into politics. Instead, he learned to connect, and his newly found focus on the audience was critical to his success.

He would always direct me to avoid dimming the lights when he spoke. Many politicians prefer a darker setting, using the spotlight as a way to focus on them and freeing the audience to settle into their chairs and remain distant and comfortable. Reagan was just the opposite. He wanted the audience to have the same amount of ownership in the event as he did. It was a marriage in his view, not an address, and like any partnership, it required a mutual commitment. He liked to see into their eyes, to gauge the effectiveness of his words and movements. Every speech was a new adventure, not just as paid company man or governor, but as president as well. He would be the first to admit it.

It was during his tenure as GE spokesman that he learned it was best always to dine with his listeners. Later in life, Richard Nixon would advise Reagan that he should eat his dinner in solitude in his hotel room where he could rehearse one last time and relish the quiet time before facing the gritty masses. This would not only provide the opportunity for singular, uninterrupted dining, Nixon advised, but would also afford a triumphant entrance into the ballroom. Reagan respectfully discarded the loner Nixon's counsel, preferring to always take his seat at the head table and dine at the same time as everybody else. He did this not out of any populist impulses but out of a need to be in the same room as his audience so he could make a personal connection before he spoke. I watched him at hundreds of these small gatherings. There would always be people lined up by his spot on the dais, wanting an autograph or simply hoping to say hello and shake his hand. Between bites, Reagan would express a deep interest in whatever occasion had brought the group together. He was always listening and learning.

The last time I saw and spoke to President Nixon was at the opening of the Reagan Library in the early 1990s. Richard and Pat Nixon came to the event despite Pat's poor health. During the ceremonies, Pat asked to relax for a few minutes to regain her strength. I led them both to a holding room and was able to spend some time alone with the former president. He marveled aloud at Reagan's ability to connect with the people. He talked about how he could never make the same connection. I nodded sympathetically. Nixon mistakenly thought it was the way Ronald Reagan modulated his voice. Although I didn't say it then, poor Nixon could never get the personal contact thing right. One story that I think best illustrates Nixon's inability to connect occurred during an interview break with the legendary David Frost. Instead of telling an off-the-cuff Reagan-like joke, Nixon asked his host, "Well, David, did you do any fornicating over the weekend?"

Being a Californian, and Republican, I relished the comparisons between Nixon and Reagan. They were so remarkably different, yet both had followed the same path to prominence. In 1972, Governor Reagan served as Nixon's California reelection committee chairman. Late one day the phone rang in Reagan's office. It was President Nixon calling from on-board the presidential yacht, *Sequoia*—and he was not pleased. I was in the room with the governor, and I could hear Nixon's voice shouting from the earpiece. He was outraged because Reagan had selected a certain congressman as his San Bernardino chairman. "Why'd you do it? He's a jerk," I could hear Nixon ask.

Reagan delicately placed his hand over the receiver and whispered to me, "Who is he talking about?" And thus the world of difference between two of California's greatest political sons. One was hysterical over who would run his campaign in a town of seventy-five thousand people in his home state. The other, the governor who appointed the man, didn't know the appointee's name. Reagan didn't pay attention to minute political details, but Nixon used to be able to name every precinct chairman in the country. Reagan knew his own shortcomings and concentrated on what he did best.

Reagan also learned before entering politics that to connect with your crowd, you need to be physically close to them. Since the lights were already on, per his direction, he wanted to be sure that his listeners were within striking distance. I was to instruct the organizers to put the first row no more than eight feet from the lectern. If the front row were any farther back, I'd hear about it later. He wanted that eye contact.

Reagan's love for intimacy with his audience and his comfort with the medium inspired me to suggest reviving the weekly Saturday morning radio address. When I first floated the idea to the president and his senior staff in 1981, I was greeted by moans and rolling eyes. Radio was a dying medium they said; no one would care. Reagan, however, seized the idea. He had done radio before entering politics. After he left the governor's office, he'd weighed an offer to do twice weekly commentaries on the *CBS Evening News* before settling on a daily five-minute radio address. He also could remember FDR's warm, inviting voice flowing into his living room as a young man. He agreed, and since then, a generation has grown up hearing at least bits and pieces of addresses from Presidents Reagan, Bush, and Clinton talking about everything from Middle East peace to school uniforms.

Judging and connecting with an audience and making them part of the action—these were the lessons that forged Ronald Reagan and served as a foundation for the development of the Great Communicator. Along with an increasingly conservative message, it was these skills that attracted the attention of the GOP financial leaders who comprised the Kitchen Cabinet in California.

In early 1965, the Kitchen Cabinet summoned veteran political operatives Stuart Spencer and his partner Bill Roberts—fresh off the Rockefeller for President campaign against Goldwater—to see what they thought of Reagan's chance against Governor Brown. The Kitchen Cabinet had already talked with Reagan and thought he was a winner, but they wanted an opinion from the political pros.

Stu Spencer had heard the same things about Reagan as the Kitchen Cabinet. He, too, listened to The Speech and found it a balm for a wounded party. He knew that Reagan had spent considerable time on the "campaign trail" as a GE spokesman, but Stu's knowledge pretty much ended there. A face-to-face meeting was needed to make a more learned assessment.

Not long ago, Stu and I got together at his home in Palm Desert to relive those days. Stu and Reagan had met at Reagan's house in Pacific Palisades. Nancy attended as well but was quiet during the meeting. After spending an hour or so with Reagan, Spencer got back to the Kitchen Cabinet, reporting that Reagan was charming, but green; motivated, but without direction. Spencer said Reagan's enthusiasm for running, a critical ingredient, was palpable. Clearly, despite Reagan's inexperience, he wanted the prize.

Reagan and Spencer then went separate ways, both making phone calls to confirm each other's veracity and capabilities. Reagan spoke at length with Goldwater who advised that Spencer was a "tough son of a bitch, but the best in the business—get him if you can." Reagan knew that he wanted Stu to run his campaign for governor.

Spencer wasn't as confident of the marriage as Reagan. He took his time getting back to the former actor, knowing that he'd be signing on to a somewhat risky campaign, in large part because many in the GOP establishment were already lining up behind Christopher. In one memorable exchange, Spencer spoke with Jack Warner of Warner Brothers Studios. When asked about Reagan's chances for becoming California's chief executive, Warner cut him off: "Wrong! It's Jimmy Stewart as governor, Reagan as best friend." Warner had been in the movie business for six decades by then; it's no surprise he couldn't step outside his world and see Reagan clearly. But Stu Spencer knew politics, and he would soon be sold.

From his many conversations, Spencer had gleaned that Reagan would have a tough time getting the nomination. Reagan was unschooled in the day-to-day minutiae of state government. He had little management experience to point to. The only office he ever sought was that of president of the Screen Actors Guild. Stu told me that when he picked up the phone a month later, it was the distinctive voice of Ronald Reagan he heard. "Stu, are you going to shit or get off the pot? I need an answer." "Ron, I'm still not convinced you're not a John Bircher," Spencer answered. He didn't really mean it, but the extreme right wingers who made up the John Birch Society were a handy excuse for his indecision. Undeterred, Reagan demanded and received another meeting.

Nancy was absent at this one, and maybe because she was, Reagan's sense of style was absent, too. Decades later, Stu still chuckles at the ex-actor's clumsy effort to promote a Birch-free, moderate image by donning groovy white pants and bright red socks. The image remains seared on the back of Spencer's brain. My guess is that Reagan simply liked the socks—he wouldn't have dressed to make that point. Still, in time, Reagan alleviated all the concerns that were raised and Spencer agreed to take the job as campaign manager. Ground rules needed to be set immediately, and Spencer remembers saying, "One, I'll always tell you the truth, no matter how unpleasant; two, I'm the boss of the entire campaign; and three, if there are any disputes within the campaign, you have the final say." Reagan reached out his hand and told him he had a deal.

On the way out, Spencer solemnly cautioned Reagan that his candidacy "might not catch on fire." He told him that nobody was doing handstands for an ex-actor. There was no groundswell, he said. Reagan nodded without emotion, tacitly acknowledging that it would be tough not only to beat the charismatic Christopher for the GOP nod, but also to unseat a sitting governor.

Of course, Spencer was not playing straight with his new client. As soon as he found the nearest pay phone, he reported back to the Kitchen Cabinet that he "just signed on with the next governor of California." It would be a relationship that would soar to dizzying heights, taking Reagan to the governor's mansion in Sacramento and later to the White House. With Spencer at the controls, Ronald Reagan the actor and paid corporate spokesman took the first steps at morphing into a political giant. His prowess on the campaign trail would catch nearly everybody, including many of his closest friends, totally off guard.

Spencer insisted that Reagan tighten his Goldwater/GE speech to about twenty minutes, or at least he tried to. Back then and for the next two decades, it would never be an easy task, no matter who undertook it. The problem wasn't the editing out—Reagan understood the need for brevity and tightness. The problem was the adding back in. Reagan was constantly field testing parables, facts, and jokes with his audiences. When the response was good, he'd slip the new material into The Speech, along with old and familiar lines he just couldn't let go, and pretty soon, the new sleeked-down twenty-minute version was taking as long to deliver as the old hour-long ver-

∽↶↶∾

sion, and as always, the crowds would be lapping it up.

Still, Stu persisted. Once he got the speech pared back, he built in a half hour of questions and answers to make each appearance just short of an hour. While Reagan had already proven himself to be a solid public speaker, he was not yet ready for down and dirty questioning about state government. Reagan understood this and tried to fool no one. He agreed that while he was learning the ropes of government, he would not try to fast-talk his way past questioners. If he didn't know an answer, he'd say so.

He also agreed to weekly sessions with Charlie Conrad, a former actor turned state assemblyman who knew the inner workings of Sacramento better than anybody. Reagan was a proud man, and I think he was comforted by the fact that Conrad had labored in the same fields that he had. They had known each other from the Screen Actors Guild, and Reagan respected Charlie's deep insight in government.

Reagan proved to be an adept pupil, and Tuesdays with Charlie became indispensable for Reagan's development.

While Reagan was learning what he had to learn, Spencer told me that he realized that to get the most out of his candidate, he'd have to talk in phrases that would inspire him. He spoke in Hollywood lingo, urging Reagan to "take his show on the road" and intimating that "if we can make it in Bakersfield or San Jose or Encino, we can make it anywhere." To the outsider, this may seem a bit odd: one adult running for elective office had enrolled in civics 101, the other, charged with making him electable, was talking like a poor man's Mack Sennett. But Reagan was no ordinary candidate, and as I would learn later, he posed unique challenges to his campaign staff.

I empathize with those who have tried to get to know Reagan more than he'd allow. It's taken people years to get to know him, if at all. But it didn't take Spencer very long to see what the problem was in 1966—Ronald Reagan was a very private and shy man. He was probably the only person other than Nancy to see this early on.

For people who lived through the Reagan presidency and listened to his lofty rhetoric and memorable soliloquies, it may be hard to accept that the Great Communicator started out as a political introvert. But on the campaign trail, he was never good at small talk, and he loathed talking about himself. Not the best combo for a budding candidate, you might think.

Ronald Reagan thought that Ronald Reagan was the most boring topic in the world. He'd never dart around a reception from important person to important person. It just wasn't his style, ever. If he had to be there at all, he'd prefer to stand in one spot and talk with the same person all night. As an expression of a healthy ego, that might have been just fine, but Stu Spencer knew that if his man wouldn't, or couldn't, engage in the sort of retail politics that requires working a room, there would be trouble ahead. Put Reagan on a stage, though, or at a lectern, or on top of a milk crate and a metamorphosis would take place. He'd talk about his agenda and bond with the crowd. Reagan wanted to connect, but he didn't want to give up his space or his privacy.

Meanwhile, another key player in Team Reagan was getting sea legs, too. Nancy Reagan became more and more interested in the process, perhaps sensing that her husband's candidacy was going to blossom and she would soon be forced into the limelight. Reagan also needed a heavy for dealing with people and staff, and Nancy would grow into the role. He was the softest touch around. Before he ran for governor, Reagan would drive up and down the state of California in his maroon Lincoln Continental, saying "of course" to anybody who would ask him to speak. He carried his little black schedule book with him and would talk to garden clubs and small town Rotary groups and just about any other group that asked. He couldn't say no. Nancy took that little schedule book away from him and told him to send speaking requests to the staff.

Reagan made no secret that he never would have succeeded politically without her. He would be especially lost without her help with staff problems. She would pay a great price for her efforts over time, and would never see the credit that she was due for helping Reagan stay with the right people. This was no Eleanor Roosevelt or Hillary Clinton. She didn't assume her strong role in politics because of a personal agenda. Her only agenda was Ronald Reagan, and there was nobody alive who better knew his strengths and weaknesses.

Nancy never involved herself in any of Ronald Reagan's campaigns or in matters of great substance, either in Sacramento or in Washington. Her real concerns were his schedule, especially how it might be affecting his strength and health, and what the press was saying. But pity the person she thought might be taking advantage of her husband or hurting his image.

In mid-1966, Taft Schreiber, a founding member of the Kitchen Cabinet and longtime agent to Reagan the actor, told Stu Spencer that "You are going to have to fire lots of people. Reagan simply can't do it." Deep down, Reagan disliked personal conflict and would avoid it like the plague. Not that he was afraid to make decisions. God knows, he would make watershed decisions of state, but personnel decisions were something that he could do without. Dismissing an incompetent aide or even dropping a consultant was anathema to Reagan. This quirk would have become a significant hindrance for Reagan if Nancy had not stepped in to fill the vacuum.

Perhaps he couldn't bear the thought of a staffer losing his job, or he was just incapable of hurting anybody. He just could not discipline or dismiss. Whatever the reason, Nancy tried to compensate for him by being on the lookout for people who didn't have her husband's interests at heart. She wouldn't do it herself but would select the right Reagan friends from the outside to help communicate the problem with the recalcitrant staffer. This outsider, often

Holmes Tuttle and even later I or Paul Laxalt, would step in and serve as the fixer.

The sad consequence is that Reagan always got to don the white hat while Nancy was portrayed as the Wicked Witch of the West. I know it wounded her deeply: she never got used to reading about her alleged heavy-handedness. History owes her one, though, because if she hadn't stepped up, Ronald Reagan would never have become governor of California let alone president of the United States. He knew this better than anyone. If you want to know more about how much she meant to him, read her book, *I Love You, Ronnie,* a collection of Reagan's letters to Nancy.

As the days passed and Spencer's team did its work, Reagan was shaping into a dream candidate, a politician worthy of the office he was seeking. He was soon able to enunciate without pause an ideology that seemed to come from deep within him and address more confidently the workings of state government.

There was a small dark cloud, however, that held the potential to haunt the candidate not only now, but also in the future: the rarely unsheathed Reagan temper.

I would see Mount Reagan blow only three or four times during my nearly thirty years knowing the man. Personally, I don't think that a little purple rage is bad. Everyone needs to unburden himself from time to time. It's human nature. Spencer was one of the few to see the Reagan rage up close. During Reagan's first campaign in 1966, at an appearance with Christopher, the frontrunner intimated that Reagan had racist tendencies.

You couldn't see his anger when he was on the stage. Reagan kept it under wraps. Unfortunately, he went to jelly backstage. Christopher had clearly gotten under his skin, and Spencer had his first glimpse at what he would call "The Rhythm Candidate." For the most part, Reagan followed the sort of predictable rhythm day in and day out that is any campaign manager's dream. This is one business where you don't want to face a lot of surprises.

But much to his chagrin, Spencer realized if an opponent questioned Reagan's honor or any of his deeply held values, it could throw Reagan totally off his game. Reagan brooded about these personal affronts, sometimes for days. He would squander precious time responding to what ideally should have been a one-day detour.

In retrospect, some of the charges that would drive him over the top would be mild when compared to today's hardball, but Reagan truly believed that his honor was beyond reproach. He would continue to protect his good name, even if it meant going "off message" for several days. Spencer took note of this rare lapse in the famed Reagan discipline and would later use it against his boss during the 1976 primaries when he was managing the Ford election drive.

The critical Midwest primaries were taking place when Spencer launched a salvo in the form of a seemingly ill-advised television ad in California. Remember: Reagan was the former governor of the state and he would handily win the primary there, but Spencer decided to make a significant ad buy in the Golden State in advance of the all-important Ohio contest.

The Ford ad simply played on the fears that many Americans had toward Reagan at the time, that he was a cowboy with an itchy trigger finger. "Governor Reagan couldn't get us into a war, but President Reagan can," the ad charged. A nervous advance man boarded the plane when we landed, pulled me aside, and handed me notes that outlined the commercial. After he left, I told Reagan what happened in California. He went into a rage slamming his fist into the bulkhead wall of the aircraft. He believed that his integrity was being questioned, and he knew it came from a close friend who knew better. "That damn Spencer's behind this," he said, accurately.

When Reagan appeared before the press in Ohio, hordes of reporters were waiting. Reagan kept his cool throughout the questions, many of them the same softies tossed to him on countless occasions. When the questioning petered out, Martin Anderson, Reagan's economist, and I escorted Reagan toward a holding room. I was just starting to relax, thanking the good Lord for getting us through the media session, but just as we turned our back, an NBC reporter asked Reagan about the California ad. Reagan spun around on one heel so fast that he literally sent poor Anderson into a wall. The rage was back. Fists didn't fly of course, but he went into one of those endless let-me-tell-you sort of sound bites that make any campaign manager's hair stand on end. The rosy cheeks and the luring smile were gone, replaced by a purple tide and a quivering lip. The scenes of Reagan pointing to the reporter were shown on local and national television almost immediately. Reagan knew he had blown it and said as much behind closed doors.

I was seeing The Rhythm Candidate up close and in person for the first time, and it was not a pretty picture. It would take us days to get the man back on track; by then the damage had been done. The famed Reagan discipline had its limits, a fault line that could be exploited by those in the know. Once more, Spencer had incurred the wrath of the Reagan crowd, just as he had in 1968 when he went to work on Nelson Rockefeller's campaign for the Republican presidential nomination. But whoever was signing his checks, Stu was always an astute strategist, and this time he had truly earned his pay from Gerald Ford.

Reagan didn't coin what became known as his Eleventh Commandment: "Thou shall not speak ill of a fellow Republican." That honor belonged to a former California state party chairman, Gaylord Parkinson, but Reagan made the commandment his own, frequently using it in those early days to defuse attacks on his lack of experience and later invoking it to soothe the waters during contentious GOP presidential primary seasons. In reality, though, Reagan would fight until the end if he believed in something. He didn't care if his opponent was a sitting GOP president like Gerald Ford, as long as personal attacks were off the table.

On the campaign trail I learned many things

from Reagan—all the things you expect to glean from a great campaigner and some you wouldn't dream of in a million years. Periodically, he would put down the memo he'd be reading on the campaign plane and start talking about something that would come totally from out of left field. One time he spent an hour telling me how and when somebody should perform a tracheotomy. I'd look at him with a bent head, curious to know where he'd picked up such a pearl of wisdom. I never found out, but my best guess is that his father-in-law, Loyal Davis, had given him this mini-lesson in tracheal surgery. Reagan was always fascinated with medical information. He never drank real coffee, he said, because it would "harm your kidneys." When he did have a cocktail, he preferred screwdrivers so he could soak up vitamin C from the orange juice.

Reagan believed God has a plan for everything, and maybe he was watching over Reagan. Early one morning in the governor's office, Reagan started to tell me about the

Heimlich maneuver, a technique to be employed when somebody is choking on food. At one point, he actually got behind me and demonstrated by squeezing me firmly below my rib cage. Thankfully, nobody noticed this future candidate for president of the United States hugging his top campaign aide. The bubba vote would have been out the door. The brief if odd lesson, though, would prove to be worth it.

Reagan was a man of routine, and a month later, after a campaign appearance, he went into his usual preflight drill. He asked the stewardess for a bag of peanuts and a Coca-Cola. He'd toss a nut in his mouth, one at a time. On this particular day, as he began his routine, I went to work on a novel. For some reason, I felt moved to put the novel down just as Reagan tossed down another nut. As the force of the takeoff pressed his head against the seat, I noticed he had a pained, discombobulated look on his face. Immediately he began to wrestle with his seat belt. Nancy, sitting next to him, shouted to the Secret Service agents, "Do something!"

Reagan had finally unfastened the seat belt and tried to stand up, but he was too tall to fully extend. The agents responded quickly, thinking he was suffering a heart attack. Over Nancy's pitched screams, one agent grabbed the oxygen, while the other tried to rip Reagan's shirt off. Reagan—turning a lovely shade of red—looked at me directly, his eyes telling me to get with the program, and I knew then what was happening. Pushing my way past the agents, I got behind Reagan and did the squeeze that he'd demonstrated for me only weeks earlier. After a few forceful shots to the abdomen, the wayward nut popped quietly off the bulkhead.

After the commotion died down, Reagan leaned over to me and said, "I'm sure glad I taught you that darn thing." The future president of the United States was almost brought down by Mr. Peanut. After Paul Harvey recounted this story on his radio show a few years later, Dr. Heimlich himself called me for details.

* * *

Reagan lived a charmed life on the campaign trail, but it was a tough place and he was older than he looked. He was warming up for the presidency when most people are in their fourth or fifth year of retirement. As I write this, I am sixty-two years old, seven years younger than Reagan when he started his second run at the presidency in 1980, and I know I'm feeling the years.

Reagan took a lot of heat for being the oldest man to ever run for president. He defused these attacks with humor. "You know, I've already lived some twenty years longer than my life expectancy was at birth," he would say, "and that has been a source of annoyance to a number of people." Or he would say that he remembered when a hot story broke when he was a young fellow. "Reporters would come rushing in yelling 'Stop the chisels!'" My favorite was a favorite of his as well: "Well, I sure know more about being young than I do about being old."

In the 1984 campaign against Walter "Fritz" Mondale, Reagan admitted that he was "flat-tened out" during the first debate against his considerably younger opponent. The age issue was back, and newspapers and pundits were wondering aloud if Reagan was up to a second term. After all, he was now seventy-three. These reports concerned Reagan, but they should have concerned his staff even more. We handled the debate preparation in 1984 the same way we had done it in 1980, forgetting that Reagan had been president for four years. Instead of equipping him with a broad list of themes and letting him fill in the facts on his own, we crammed him with details and ignored the big picture. As I sat there watching him that first debate, I could see his mental computer was so jammed with facts that it kept short-circuiting. We hadn't let Reagan be Reagan, and he paid the price.

In the second debate, Reagan would wipe away the doubts about his age in one fell swoop as he responded to a question from Henry Trewhitt of the *Baltimore Sun,* who asked Reagan about his ability to stay up all night if a national crisis were at hand. "Is there any doubt in your mind that you would be able to function in such a circumstance?" Trewhitt asked. Candidate Reagan answered very calmly and with an even, serious tone. "Not at all Mr. Trewhitt, and I want you to know that also I will not make age an issue of this campaign. I am not going to exploit, for political purposes, my opponent's youth and inexperience." Everybody cracked up, including Fritz Mondale. From then on, the age issue seemed to disappear as a major concern. For the rest of the evening, Reagan was his old self.

Reagan usually came up with these zingers on his own. He planned on using the line against Mondale, but most of his best material was totally off-the-cuff. Remember "There you go again" against Carter in 1980? It single-handedly turned the campaign around and humanized Governor Reagan with a previously skeptical electorate, and it was entirely home-grown. He came up with it on the spot.

Reagan regularly used humor as a weapon in his campaigns concerning other issues as well. In 1980, taking advantage of the economic malaise suffered under the Carter administration, Reagan prescribed: "A recession is when your neighbor loses his job. A depression is when you lose yours; and recovering is when Jimmy Carter loses his."

Candidate Reagan would poke fun at the federal bureaucracy, using its excesses as a reason to vote for him. Later he would often recount a fictitious yarn of a sobbing bureaucrat he encountered at the Bureau of Indian Affairs. The man was at his desk, crying into his folded arms when Reagan touched him on the shoulder and asked him what was wrong. "My Indian died, that's what's wrong," came the response. "What the hell am I supposed to do now?"

On the trail and later in the White House, he'd always have a store of good jokes to put his visitors, friends, and staffers at ease. The rigors of campaign life would get to just about everybody, but Reagan handled the many pressures with ease. Not that he was above a gaffe or two, but he did a good job of keeping them to a minimum.

One day during a campaign stop in an Ohio steel town, a reporter was hectoring Reagan on his policy regarding clean air. Reagan started going down a slippery slope, eventually claiming that trees themselves contribute to air pollution. The brief statement would come back to haunt the candidate many times over. The most entertaining retribution occurred when we flew into smoggy Burbank. As the motorcade rolled toward another campaign stop, Reagan and I noticed a large sign hammered to a eucalyptus tree: "Cut me down before I kill again." Reagan joined the laughter without conceding that he'd been off target.

Even though I had worked with Reagan for nearly fifteen years when he ran for president again in 1980, I was still looking for new ways to let the American people see him as I did. I was convinced that if voters could get an unvarnished look at Reagan, they'd pull the lever for him every time. I think Reagan believed this, too. He knew that if he didn't get a connection with his audience—be it in person or over the airwaves—he'd be less effective. This is part of the reason why he never let makeup touch his face. Never. Like a lot of others, I'd been fooled the first time I saw him in person. The guy was Hollywood, I figured—he'd do anything to look better—but nothing could be further from the truth.

Nature would sometimes betray Reagan as he got older, and he wouldn't always have the fresh, rosy look I first saw at the Ambassador Hotel in 1964. Happily, as I learned, there was a natural remedy close at hand. One evening, after practicing for a debate with Jimmy Carter, I joined Reagan for dinner at our campaign headquarters in Middleburg, Virginia. I had noticed during the mock debate that Reagan didn't have his usual glow. But during dinner, after a glass of a fine French red, Reagan seemed to get his color back. His eyes sparkled, and his cheeks turned red.

This gave me an idea. Before the predebate dinner, I left a bottle of the best French wine out on the table. Being a wine lover—one glass only, of course—Reagan would inspect it. If the year and vintage passed muster with him, he was sure to pour himself a glass while reviewing his notes. And so he did. Immediately, as if on cue, the capillaries rose to the occasion, and the youthful Reagan was back. The wine strategy was my insurance policy. Just in case he happened to be off his game, we'd have a little color on our side. It worked like a charm. Reagan unwittingly signed on to my plan and would go on to have a glass before some of the biggest television speeches of his life.

Unbeknownst to most, Reagan had a very unusual way of reading his speeches. Ever since I joined him, I noticed that he always carried two items with him when he was doing a speech. The first were his famous three-by-five-inch cards, the ink still moist from his most recent additions. The second essential item was his contact lens case. Reagan had terrible eyesight, and he needed reading glasses to see his speech when he wore his contacts. Without the lenses and the reading glasses, he would be able to read his speech but he wouldn't see his audience at all. That left Reagan with a couple of choices. One, he could read the speech without glasses or contacts and not see his audience. Two, he could wear his contacts but use reading glasses to see his speech. Ronald Reagan didn't like these two choices, so he came up with a third option. He would simply remove his left contact and keep his right one in. This way, he could read his speech with his left eye, but still see his audience with the right. Heaven knows how he was able to perfect this, but I saw him do it a million times. As soon as the speech ended, he'd pop the other "eye" back in.

Was it vanity that drove him to this odd arrangement? I don't think so. Voters are moved in large part by visuals. What would the people have said if they'd seen the sixty-nine-year-old Ronald Reagan wearing reading glasses in his first debate with Jimmy Carter? It would have reinforced every stereotype people had about him at the time. I was always hypersensitive to the way Reagan looked and how he came across to the average American. If you could summarize my role with Reagan for all those years, I guess you could say I was the guy who helped him look good, but, really, all I ever did was light him well.

One of the more difficult obstacles I had to get past, believe it or not, was that Reagan didn't like to have his picture taken. Before the 1976 election cycle heated up, we scheduled a photo session so we could get the definitive Reagan campaign photo. This picture would be used in countless posters, buttons, and flyers.

We had squeezed the photo session into an afternoon of campaigning during the Florida primary. When the set had been lit and everything was in order, I went to Reagan's door to bring him to the small studio. To my horror, I found him wearing the most awful tie I had ever laid eyes on. It was bright orange, with large dark black blots all over it as if someone had used it for a makeshift Rorschach test. Other than that, he looked perfect. The photographer waved Reagan over, indicating that he was ready to take the first round of shots. I pulled Reagan aside and asked if he'd like to select another tie.

"Why would I want to get another tie?" he asked, puzzled and frowning.

"Well, Governor, for a shoot like this you're going to want to use a tie like mine," I said, offering it to him. "It's got a more subtle pattern, and it'll show up better on film."

"I don't like your damn tie, either," he replied curtly.

I went on to tell him that this picture would be replicated millions and millions of times. It would be the only picture we'd be taking for our campaign materials. He looked at his tie, then at mine, and walked out of the room. He appeared ten minutes later, frowning, with another tie. By now, he was in such a terrible mood that we never did get a good shoot. Today, we could have just taken the photo with the bad tie and added a new one using a computer.

Later I told him that I noticed his aversion to sitting for photo shoots. He looked at me surprised. "That's funny, in all these years, no-body's ever noticed that." I asked him to elaborate. "Well, you can never recover from a still shot."

Reagan was most comfortable with moving film, he went on to say. He truly believed the television camera was a friend, a device that would separate the real from the phony. Still cameras could always be used to make a candidate look like a fool. When he explained this to me in the late 1960s, he said, "You know how I sometimes touch my nose before I make a point? Well, a still shot would show me picking my nose, while a live shot would show me making my point."

I recall watching old movies with him, constantly asking what the actors were really like in person. I once told him I thought Marlon Brando was a jerk based on the way he felt to me on the screen. Reagan didn't say anything.

"Well, was he?" I demanded.

"Mike, remember, the camera never lies."

Conversation over, I had my answer and would never forget it. Imagine if every candidate understood that he couldn't fool the camera.

Sometimes, if we're lucky, age brings a little wisdom, and so, I hope, it has been with me. Back when I was in the thick of Reagan's gubernatorial and presidential races, I could have ticked off in a second the top five reasons people voted for him. Now I know that Reagan was always a hundred times a hundred different things to the people who voted for him. His wide appeal hit me with special force one night in Iowa before the 1980 election. We were barnstorming the Midwest and were still slightly behind Carter in the polls. Throughout the day, I kept getting notes from our advance men in the field that former senator Eugene McCarthy, one of the most respected liberals to ever serve in the Senate, wanted a private meeting with Reagan as soon as possible.

I couldn't understand why this icon of the left would want to meet with the new Mr. Conservative, especially during a heated campaign. When we got to our hotel, there was

Eugene McCarthy, sitting by himself in the lobby. Judging by the stack of exhausted newspapers on the floor next to him, I guessed he'd been there for a little while.

I felt guilty not getting back to him all day, but then a rush of panic came over me when McCarthy said, "Governor, I'd like to talk to you in private, if that's okay." Reagan nodded and shook his hand. Instinctively, I followed the two men into Reagan's room. McCarthy didn't seem to mind, so I hung around long enough to hear him say that he wanted to formally endorse Ronald Reagan for president. McCarthy outlined his reasons for supporting Reagan and abandoning Carter. Governor Reagan was very appreciative, and after some small talk, I offered to walk Senator McCarthy out to his car.

As we walked silently through the crisp air, I told him that I didn't get it. "What you said in there makes some sense, Senator," I said, "but what's the real reason you're endorsing Ronald Reagan?"

As he grabbed the door handle of his car, he

looked me directly in the eye and paused. "I'll tell you why," he said without smiling but loud enough to make sure I took it all in. "It's because he is the only man since Harry Truman who won't confuse the job with the man." McCarthy knew that Reagan understood the enormity of the office he sought, that it was bigger than any one man. McCarthy saw something in the former actor that millions of Americans would soon see on Election Day as they happily handed him the keys to the White House.

CHAPTER THREE

∾

MR. PRESIDENT

I was with Ronald Reagan the moment he knew he was going to be the president of the United States. It was Election Day 1980, and at the Reagans' Palisades home veteran CBS anchor Walter Cronkite called to tell Governor Reagan that he was about to call the election for him. I looked over at the sixty-nine-year-old president-elect. He was not overcome with emotion. He simply rose, kissed his wife on the lips, and looked at me with a discernible twinkle in his eye and a satisfied look on his face. Several hours later, he was in the shower when Jimmy Carter called to concede. With no time to do otherwise, Reagan took the call dripping

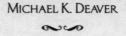
wet, with a towel around him. Carter had been most gracious, Reagan told me afterward, touched by the president's earlier-than-usual concession.

That evening Carolyn and I were at the home of Earl and Marion Jorgenson for an election night dinner, which had become a ritual for the Reagans. The newly elected president and First Lady were joined by the Kitchen Cabinet, top aides, and, of course, old Hollywood friends. I remember spending time with Jimmy and Gloria Stewart that night in 1980. Jimmy had campaigned vigorously for Reagan in 1976 when few celebrities would stick their necks out. He was a great asset with a great sense of humor. Once on the campaign trail, during a windy airport rally, Stewart's toupee had been rising noticeably up and down with the gusts, crouching like a small cat. After he got back on the plane, he stuttered in classic Stewart-speak that he would have to get some "t-t-t-tacks if we d-d-did that again."

Later, during the Jorgenson party, I found Reagan's brother, Neil, alone in a room watching his brother on the television. Tears were streaming down his face. "This must be a great moment for you," I said. Between sobs he managed to answer, "I only wish that our mother Nelle could be here to see it."

Ronald Reagan's rise from actor to governor to president was a roller-coaster ride that changed everybody except the guy in the front car. I was with him not only when the election returns came in, but also at the precise moment he decided that he wanted the big job. With Reagan, there were no senior staff meetings to discuss such stuff, no trial balloons floated on the evening news to gauge the public's interest. Sure, there had been countless meetings where Reagan would listen, but he never said, "I'm running" until we were sitting together on an airplane in early 1976. Reagan had left the governor's mansion just two years earlier, and he'd been wowing conservative groups across the country with The Speech ever since. We had just taken our seats on a flight from Los Angeles to San Francisco when an attractive woman he had never met before knelt in the aisle next to his seat. "Governor," she said, "you must run for president. You must do it for the people who believe in the things you do."

I laughed to myself. I had seen it a million times before. Reagan would grin and shake her hand, thank her, and maybe sign an autograph if asked. This time, though, he just nodded and told her to have a good day. After she took her seat a couple of rows behind us, Reagan looked to see if she was out of hearing range. "You know, she's right," he said in a hushed tone. "I don't think Jerry Ford can win, and if I don't run, I'm going to be like the guy who always sat on the bench and never got in the game." I remained quiet. It was my policy not to bring the presidency up. I always believed that the fire in the belly needed to come from within Reagan, not from the anxious people around him.

I listened for the next hour as he went over the pros and cons of taking the plunge. Ten different men would probably cite ten different reasons to run for president, but Reagan simply knew that he could, unlike Ford or Nixon, connect with the people. This was the key, he said. He didn't use sophisticated polling or do an Electoral College breakdown. He didn't care if he lined up all the GOP governors, senators, or congressmen or how much money he'd have to find to run a national campaign. I think he just believed in his heart that he was the right guy at the right time.

Toward the end of that memorable flight he explained, "Mike, I remember in the movie *Santa Fe Trail*, I played George Custer as a young lieutenant. The dying captain said to me, 'You have got to take over.' And my line was 'I can't, I can't.' And the captain said, 'You must, it's your duty.' That's the way I feel about this, it's my duty," he said. "I have to run. I'm going to run." Reagan would often use movies as ways to illustrate his logic. To me this made perfect sense—that's the world he knew—but I also knew this time that he was dead serious.

After the election victory, we had just two months to make the transition from cam-

paigner to president. Election Day to Inaugural
Day was a blur for me personally. I hadn't been
so sure I wanted to move east to help with governing. I also wondered how Reagan would
handle the pressure once he was formally installed in the Oval Office. He had never wavered before, but this was an entirely new set
of pressures.

Ronald and Nancy Reagan stayed in Blair
House, the presidential guesthouse on Pennsylvania Avenue across from the White House,
during the awkward period leading up to the
swearing in. I say awkward because as the Reagans prepared to move across the street, the
current occupant of 1600 Pennsylvania Avenue
remained entrenched inside. The embassy staff
in Tehran had been taken hostage on his watch,
and understandably, Carter was prepared to
fight to the last minute to make sure they tasted
freedom while he was still the commander in
chief.

The morning of the transfer of power, the
traditional swearing in on the steps of the U.S.
Capitol, I arrived at Blair House at 8:00 A.M.

sharp to greet the Reagans. On my way over, I assumed that the president-elect would be practicing the most important speech of his life. I walked in and greeted Nancy.

"Where's the governor?" I asked.

Careful not to disrupt the labors of her stylist, Nancy moved only her mouth, saying, "I guess he's still in bed."

I didn't believe it. Although Reagan had been elected governor of California twice and nearly beat Gerald Ford in 1976 for the Republican nomination, this was still big stuff. I checked for myself, cracking the door to Reagan's room. It was pitch black. All I could hear was the quiet snores of a sleeping man. As he breathed, the covered mound moved up and down in a peaceful rhythm.

"Governor?" I said, too loudly.

A grunt, and a repositioning of the pillow, then, "Huh?"

"It's eight o'clock," I protested.

"Yeah?"

"Well, you're going to be inaugurated in a few hours as the fortieth president," I advised.

"Do I have to?"

If I had ever doubted Reagan's ability to handle pressure, I could have doubted it no longer. Here was a man very confident of his ability to cope with whatever lay ahead.

I accompanied the Reagans to the White House Blue Room where the customary coffee was to take place with the incoming and outgoing residents. I sat with the extended Reagan family and his closest advisers, many of whom, like me, were from California. Few, if any, of the Carter inner circle were there. Most, I assume, had packed up and headed home. The White House staff consisted of the usher and butlers who served juice and coffee. The official escort from the Senate was on hand as well.

After a brief but somewhat disconcerting delay, President Carter arrived. He was a wreck— pale, even ashen. He must have been working the phones all night, still trying to slay the dragon that haunted his presidency. The mood in the Blue Room was cordial, but just barely. Carter was so determined to finish the job at hand he barely said anything to Reagan. For his part, Reagan treated the sitting president of the United States the same way he acted around the coffee crowd at the Bel Air Presbyterian Church after Sunday services. It was one of the first times that I realized the consistency of Reagan's character. All of us were excited at our new lives, eyeballing the stunning Blue Room, dazzled to be in one of the world's great institutions. The Reagan people were all wondering how we got there and if we were up to the job of governing. Not Reagan. He was the same now as he had always been: confident, but quiet; standing erect among the various aides, butlers, and the soon-to-be-ex-president.

Nancy Reagan and Rosalynn Carter were cordial to each other, certainly, but I found it hard to tell what the outgoing First Lady was feeling about her husband's predicament, the transfer of power, or the White House's soon-to-be new occupants.

When the coffee ended, the president and president-elect, as is the custom, drove up to Capitol Hill together. They were alone in the car, but Reagan later told me, Carter remained on the phone with his National Security staff nearly the entire time.

When we got to the Capitol, Reagan pulled me aside. Instead of feeling slighted, he was deeply moved by Carter's determination to see our people free. Reagan himself was becoming increasingly agitated that a handful of Iranian zealots could bring a sitting American president to his knees. He genuinely felt for Carter and understood entirely what was going through his mind.

Later, in the holding room, Reagan waved me over.

"What is it, Governor?" I said, calling him by that title for the last time in my life.

Based on the intelligence information he was provided, Reagan thought the release of the hostages was imminent. "If it happens, even during my Inaugural Address, I want you to tell me. Slip me a note. Interrupt me. Because if it happens, I want to bring Carter up to the platform. No country should embarrass and humiliate any president of the United States."

The opportunity never came. Twenty minutes after Reagan took the oath of office, word came to us that the hostages were in fact free. At the time, the new president was having a traditional lunch with the Senate leadership. I wrote out a quick note and made my way toward President Reagan. Without looking at the note, he instinctively knew what happened. He read it and let out a long, silent breath and smiled. America had a clean slate—and a new president. He knew that the Iranians did this to embarrass Carter. You would never hear Reagan boast that he had anything to do with the release of the hostages.

An hour later, Reagan stood viewing the honor guards from the four branches of the armed services. I could tell he was relishing the moment, standing proudly next to a lieutenant general during the troop review. Naturally, the young men were saluting their new commander in chief as they filed by. The general to Reagan's right returned an equally crisp series of salutes. Reagan watched him closely as I stood quietly behind them on the observation platform; then he whispered to the general, "Is it appropriate for me to return their salute?"

"It is only appropriate, sir, if your head is covered," the general replied tersely.

Reagan nodded and tentatively placed his hand back over his heart—a tepid offering, Reagan must have thought, for the saluting airmen, soldiers, marines, and sailors. During a pause, Reagan leaned back to me. "I feel very uncomfortable just standing here, not acknowledging their salutes." He looked in the general's direction and pursed his lips.

I leaned in his other ear. "You can do whatever the hell you want—you're the commander in chief."

Reagan's eyes lit up. After a deliberate pause, he raised his hand slowly to his forehead, then completed a salute with such perfection that many military men would blush with envy. He did it with such zeal, I think he even made the attending general proud. I later asked Reagan how he had the salute down so fast.

"Come on, Mike, you know what they tell you," he said, remembering his days in the Army Air Corps. Then assuming a stern, straight position, he methodically walked me through the salute drill. "You bring it up like honey and shake it off like shit." For the next eight years, he'd unsheathe the salute for any four-star general and private he could find.

Ronald Reagan left the Inaugural stand around 4:30 that afternoon, headed for his new home on Pennsylvania Avenue. Nancy went immediately to the residence quarters to begin preparing for the more than a dozen balls and parties to be held that night, but Reagan said he wanted to have a look at his new work space. I was alone with him as he walked into the Oval Office for the first time as president. I stood silently by the door as he walked around, slowly at first, but then more quickly as he moved toward his new desk. The desk had been used by each president since John F. Kennedy a generation earlier; now, because of Reagan's height, an extra inch and a half had to be added to the legs. President Carter had left a heartfelt handwritten note in the top drawer, wishing his successor well on his first day. Reagan put the note down, sat in the chair, and looked at me wistfully. "Do you have goose bumps? Because I sure do."

He must have seen it in my eyes. The man had finally won the prize. After all our years together, all the late hours and long trips, I was as overwhelmed by emotion as he was. Here Ronald Reagan was, finally taking possession of one of the most famous rooms in the world—a place where giants like Jefferson, Lincoln, and Roosevelt had worked—and I was beside him. Thoughts of grandeur must have been racing through my mind because the president brought me back down from the clouds with a simple question. "By the way," he asked, as he rummaged through the desk drawers, "where are you going to be sitting? I hope you're close by."

"I don't know, Baker and Meese have this place pretty well carved up," I answered, referring to new Chief of Staff Jim Baker and Counselor to the President Ed Meese. "Don't worry," I added. "I'm sure they'll find me an office tomorrow."

The president stood up and walked to his left, opening a semi-hidden door that leads to a private hallway and a small but lavishly appointed office—complete with fireplace—that serves as the president's private study. Presidents Eisenhower, Truman, Kennedy, and Carter had used that room to do the nuts and bolts of the job, reserving the Oval Office as a ceremonial space to do the nation's business.

"Mike, how about this?" Reagan said pointing into the office.

I was flattered of course, but thought he was being a bit hasty. "Mr. President, you're going to want to use that room," I said. "Think it over a bit."

"Mike, I've been trying to get in that nice big round office for years," he said, hoisting his extended thumb back through the door. "Why the hell would I want this little square one?" It was a telling moment. I never thought Reagan pined for the presidency, but he treasured this most satisfying of moments.

For me, of course, the president's "gift" of this precious piece of Washington real estate was a godsend. My days began around 6:30 with breakfast in the White House with my assistant. By 9:30 in the morning, I had usually met with my own staff, Jim Baker and Ed Meese, the senior staff, and the president, both alone and with Ed, Jim, and Vice President George Bush to go over national security matters. From there, the meetings—with Reagan, the cabinet, the communications people, and so on—would stretch out until about 7:30 in the evening, with the possibility of official functions after that. I needed a place to get away, and the study was great for that. A beautiful Childe Hassam flag painting hung on the wall when I first saw the place—it had been a favorite of Jack Kennedy's and would later be admired by Bill Clinton. When the White House curator found out that I was going to be using the office, not the president, he sent word that the painting had to be put back in storage, but I struck a deal with him, and it hung there the whole time I was in residence. Even better, the study looked out on a private terrace, with a huge walnut tree just beyond. In good weather, Reagan loved to lunch out there: the fresh air restored him. And, of course, this truly was a private office. When international crises arose, I'd often hear a knock on the door—Ronald Reagan would never just barge in—and the most powerful man in the world would slip in to follow events on CNN while the National Security Council struggled to catch up.

After I left, the president turned my office into a secluded dining room, but his love and reverence for that "nice big round office" would be unflinching. He understood fully the gravity of the position he was in and the trust that the American people had bestowed on him. He knew he was not just a steward of the presidency, but also of the physical office itself. Little things showed how much he respected the institution.

One such moment came the following summer when the president and his domestic policy team were conducting a ceremony in the Rose Garden. It was one of the hottest days during one of Washington's soggiest summers—an event tailor-made for wilting the staff, including me. After the session ended, I followed Reagan back into the Oval Office to complete a discussion we were having prior to the Rose Garden ceremony. As soon as we entered the office, I took my coat off and began tugging at my drenched shirt, trying in vain to cool down. It was a fool's errand. I looked at Reagan, who had already taken his seat behind the desk. He was searching for his favorite fountain pen. His suit jacket remained on, his shirt still looked quite crisp. If he was as baked as I was, he tried not to show it.

"Mr. President, why don't you take your jacket off before we start," I suggested. "You'll be much more comfortable."

"Oh no," he said immediately, shaking his head slowly and waving his right hand in the air. "I could never take my coat off in this office."

Reagan's jacket would stay on through that get-together and through myriad mind-numbing budget meetings, security briefings, and even impromptu surprise parties.

As much as Reagan relished the job, I knew there were times when the pressures of state would force themselves even on him. One Saturday morning in the spring of 1983, I was upstairs in the residence, looking for the president, when a steward directed me toward the family dining room. The president was alone, looking out the window down Sixteenth Street. He looked pensive, almost sad.

"Penny for your thoughts," I blurted.

Knowing my voice, he didn't need to look over. He kept staring down the busy boulevard. Then he smiled reflectively and said, "I was just thinking. You know, it would be great to go poke around a bookstore or do some window shopping today."

I didn't say anything, unable to grasp the co-nundrum of being a prisoner in a palace.

"I can't even build a fire," he continued. "If I'd try, one of those White House guys would come in and say they have to do it for me."

People forget that Reagan was a true man of the West. He was no armchair cowboy; he sa-vored time on his ranch using his hands, doing things that many people leave to minimum-wage laborers. After a couple of days on his beloved Rancho del Cielo, a dozen miles north of Santa Barbara in the Santa Ynez mountains, he would look younger and stronger; the land was a kind of spiritual tonic for him. Whenever we crossed the Rockies in Air Force One head-ing west, he would always point out his win-dow and say, "The sky's getting bigger every minute," and you could feel him expanding as well.

One of my primary duties was to manage the White House schedule. I'd always make sure to carve out significant ranch time for Reagan to look forward to. That way, when he was feeling overburdened, I could point to the scheduled time at Rancho del Cielo as a way to refresh him. One year, I directed William Sadleir, my colleague who managed the president's schedule, to block three weeks in the dog days of August. Faced with what seemed pressing political issues, I later pared the ranch time back to sixteen days. Reagan saw the reduction and summoned me into the Oval Office.

"Mike, I noticed that you made some changes on the schedule," Reagan said pointing to a copy of the master calendar, the six-month schedule that I updated weekly. "Now, I usually let you fellas decide about the schedule, but I'm going to let you in on a secret. I firmly believe that the more time I spend at the ranch, the longer I am going to live," he said. "So let's keep that in mind, okay."

I waited for him to start laughing or at least smile. He did neither. Needless to say, I pushed back the "pressing business." Presidents aren't like a lot of other chief executives. They can't go sneaking off to their country retreats when the burdens of the office get overwhelming.

Reagan needed his ranch outlet, and once he got there, getting him to leave early could be ugly.

The president was at Rancho del Cielo the week before Labor Day 1983 when MiG fighter planes shot down in cold blood Korean Airlines flight KE007, which had veered innocently into Soviet airspace. A total of 269 persons, including 61 Americans, perished. After National Security Adviser Bill Clark briefed Reagan in person on the barbarous incident, I called the ranch from Lake Tahoe, where Carolyn and I were vacationing with the kids. The boss was staying put, I was told. He'd handle things from California. So far so good, I thought—Reagan had been quietly letting the farcical Soviet denials wilt under their own feeble weight—but I wasn't convinced by the strategy. Worried, I had the president tracked down in the tack room, out in the barn.

The phone was picked up hastily. "Yeah, Mike, what do you want?"

"I think we need to go back to Washington," I said.

"Why?" he said brusquely.

"Because you're the president, sir, and this is a worldwide crisis," I replied.

"Dammit, I'm the president whether I'm in California or the Oval Office," he said.

"I'm sorry, but the American people expect their leader to be in Washington in times of crisis," I said. "It's just the way it is."

"Well, it's a stupid idea if you ask me," he said, slamming the phone down.

The ride back on Air Force One was a quiet one for the first hour. Reagan sat alone, looking out the window, reflective and detached. With the president still in his ranch garb, we went straight to the situation room and listened to nearly ninety minutes of America's best and brightest foreign policy minds express their recommendations for various levels of U.S. actions. Reagan didn't say a word during the entire array of presentations from the secretaries of state and defense, the joint chiefs, and the National Security Council. They served up a helping of red meat, offering wide-ranging sanctions, an increased defense posture, and other ways to kick the Soviets when they were down. Reagan let them fight it out for a while; then he clicked his pencil on the table a few times, using it as a makeshift gavel.

"Fellas," he interrupted, "I don't think we need to do a damn thing."

A gray cloud suddenly descended over the room. The collective counsel was being discarded like yesterday's newspaper.

"The entire world will rightly and vigorously condemn the Soviets for this barbarism," he said. "We need to remember our long-term objectives." Maybe because he had been an actor, Reagan's sense of timing was exquisite in matters large and small. He knew exactly when to talk and when to keep quiet. He also instinctively knew the Soviet empire was morally corrupt, and this was yet another painful reminder why. In his mind, nothing he might have said publicly would have added to this perception.

Later that weekend, Reagan would eloquently denounce the Soviet aggression, but he would stop there. There would be no recall of our ambassadors, no increased military posture. The supposedly "trigger-happy" Reagan was the only one who wanted to do essentially nothing. He would weather conservative criticism for a month over his reticence. Once again, the different drummer had gone in a direction that few could understand.

During Reagan's first term as governor, I got an early glimpse of how he handled pressure. This was 1968. Abortion, the electric chair, taxes—big decisions weighed on him constantly. One afternoon I quietly pushed open the door to his office in time to see him take out a small bottle and place it to his lips. He took a pair of speedy gulps.

He noticed me and held the bottle up near his face so I could see it. It wasn't booze but instead a tonic for upset stomach.

"My stomach's been killing me," he offered.

This concerned me. A few months later, in the car, I decided to prod him.

"Governor, haven't seen that little bottle around lately. How's your stomach?" I asked.

Instinctively knowing that I knew why he had sought the balm, he looked at me. "Mike, when things were getting rough a few months ago, I kept looking over my shoulder for answers." He made a slow head-shaking measure, then looked at me again.

"Where I should have been looking was up," he said, pointing his finger skyward to his Maker. "That's where I found the right answers."

He took this philosophy into the White House, and a good thing it was, too, because trouble can start there on a moment's notice, from something as small as a presidential checkmark or a small note scrawled in the margins of a mind-numbing memo.

One day, Reagan scratched the words "I think we should" next to a question regarding whether the federal government should take the side of a small southern religious school in its case before the Supreme Court. The school was Bob Jones University.

Today, most people are well aware of BJU and the trouble it generously heaped upon the campaign of George W. Bush during the 2000 Republican primaries. Young Bush, though, wasn't the first big-name Republican to get tarred by the school's eighteenth-century views on interracial dating and its anti-Catholicism. That "honor" was ours. At the time, BJU was a not very well known institution seeking administration support in a lawsuit to obtain a federal tax exemption that the IRS had denied.

Based on the information presented to Reagan, it looked like a no-brainer. There was little discussion of the school's segregationist policies—the reason the IRS had said no. Nor was there talk about how mixed dating was against the rules. The issue was government control. On paper, it simply appeared that a small Christian school was being strong-armed by the feds. Furthermore, the school had powerful supporters on Capitol Hill.

The civil rights community learned of Reagan's position and naturally started pressuring the White House to reverse itself immediately. I can say unequivocally that Ronald Reagan does not have a racist bone in his entire body.

True, he was a product of the Hollywood Old School, white almost to the core, but it's hard to blame someone for the times and culture they lived in. Stories from Reagan's youth show him to be well ahead of his generation in terms of race relations and sensitivity toward the plight of minorities. In his college football days, a couple of his black teammates were rejected from a local hotel. The white kids could stay, said the lodger, but not the black ones. This didn't make sense to Reagan. He could never understand why somebody would hate another person just because he had different skin color. Pigment simply didn't matter to him. So they all went and stayed at his parents' house.

If you can blame Reagan for anything, you could probably say he might have been a bit naïve on race, and for the life of him, he couldn't believe what the clamor was about the "Bob Jones thing," as he called it.

I tried to point out that the black community is ultra-sensitive to issues like these and that the administration was being perceived as disinterested in their views. Being a stubborn fellow, Reagan was unmoved, so I decided to play hardball to get my point across. Clearly, I had forgotten Nancy Reagan's advice about not using politics as an arguing point, but I was determined on this one. I asked Thaddeus A. Garrett, an African-American serving on the vice president's staff, to give me a candid assessment of what the community was saying in the churches about Ronald Reagan in light of the Bob Jones case. Another African-American, Mel Bradley, of the White House staff, joined him in my office to discuss the situation. I didn't know Thad well at the time, but to his credit, he provided me with an unvarnished view of the current black thinking. I got the information I needed, and so would Reagan.

I brought Garrett, Bradley, and a third African-American staffer, Steve Rhodes, into the Oval Office and directed them to tell the president what they had told me. People understandably soften their tone when talking with the president in the Oval Office. I've seen it happen a million times, and this was no exception. Reagan sat listening, waiting for some sort of new insight. He wasn't getting it.

I stood up. "Thad, damn it, tell him exactly what you told me. What are they calling him in the churches?"

"Sir," he said. "They are calling you the Devil and a snake."

Reagan turned pale. He looked as if somebody had kneed him in the groin. He had withstood stinging criticism over the Bob Jones decision in the *New York Times* and the *Washington Post*—that was part of the cost of the job—but seeing African-American members of his own staff ashamed of his stance and hurt by it was more than he could or would take. He called in Ed Meese and William French Smith to say that he wanted to take another look at the matter.

Later, when he reversed his position, Reagan typically refused to blame staff even though he had been ill-served by them. "The buck stops at my desk," he said. "I'm the originator of the whole thing." In politics maybe even more so than in life as a whole, stand-up guys are way too rare.

I always felt it was my responsibility to tell Reagan when I thought he was wrong. My relationship with the president allowed me this luxury although it was something I would do only in private. The closest I ever came to resigning because of a policy decision came when we did not react as quickly as we should have during the bombings of Beirut, Lebanon.

The history of modern Lebanon is an ugly one. To understand what was happening in 1983, you could go back nearly two decades to 1967, when Israel occupied the West Bank, or you could go back only seven years, to 1976, when Lebanon was torn by civil war and Syrian president Hafez al-Assad sent in his troops. What began as an Arab peacekeeping effort turned into a permanent Arab occupation. By 1978, Israel had declared a security strip in the southern part of Lebanon and aligned itself with the Christian militia. There were splinter groups supporting, or supported by, Iran and

Syria. It was a confusing mess to anybody who doesn't monitor the region for a living. Of course, the Palestine Liberation Organization (PLO) was front and center in this drama.

I won't try to recite the history of the region or the violence that took place when 241 U.S. Marines were killed by a suicide truck bomber. But Lebanon was a nightmare waiting to happen. The elements that drew the United States into the bloodbath began with what seemed to be another of those predictable flare-ups on that turbulent border that joins Israel and Lebanon.

In response to continuing attacks on its northern border, Israel had sent tanks rolling into Lebanon in June 1982. The goal of the Israeli Defense Ministry was to secure a buffer zone free of PLO influence for twenty-five miles into Lebanon. But against surprisingly timid resistance, Israel saw an opportunity to drive the PLO completely out of Lebanon. Israeli warplanes hammered Syrian missile sites in the Bekaa Valley. Its tanks and gunboats shelled PLO bunkers and hideouts in West

Beirut. While a stunned world marveled at the speed of the operation, Israel surrounded Beirut.

The Israeli defense chief, Ariel Sharon, was on the cusp of completing his dream of eliminating the PLO. But the cost to finish the job wouldn't be cheap. PLO forces had dug in behind civilian lines, mingling with the innocent population of Beirut and threatening to take the city down with them.

Like Ariel Sharon and all of Israel under the political leadership of Menachem Begin, the Reagan White House faced a difficult question. I didn't always sit in on policy briefings, but for weeks I sat there listening as the National Security Council aides kept telling Reagan that Israel could actually win if Sharon's tanks and infantry could clean out the terrorists once and for all. At the same time, Israel would drive Syria's forces back over its borders, too. To me, this strategy risked a Middle East nightmare of biblical proportions. World War III scenarios were actually taken seriously. Was it worth the risk? Could we continue to denounce Israel by

day and encourage it by night, even as innocent Lebanese casualties mounted?

The National Security Council met with the
president at 9:30 every morning, usually for a
half hour or so. On this particular day in 1983, I
sat in my office in solitude, choosing not to at-
tend the session. For the past week, I had seen
pictures of the bodies piling up in Lebanon,
and now my mind couldn't escape those images of the carnage. Finally, I rose and walked
through my door into the Oval Office.

Reagan was alone, reviewing a document.
He nodded to me but didn't speak.

"Mr. President, I have to leave," I said.

"Mike, what are you talking about?" he said
sharply.

"I can't be part of this anymore—the bombings, the killing of children."

He listened intently, but remained silent.

"You're the one person who can stop it," I
continued. "All you have to do is tell Begin you
want it stopped."

We looked at each other for a few moments.
Reagan stared down at his lap. I felt that he

was seeing the same images that were haunting me. We didn't need to say anything. After all these years together, his body language and facial expression told me we were on common ground. Reagan picked up the phone and asked his secretary to get Menachem Begin on the phone immediately.

Reagan was completely committed to Israel, but he and Begin had not particularly hit it off when they met face-to-face, especially after Begin went straight from the White House to Capitol Hill to lobby for more money for the Jewish state after telling the president that he would accept the foreign aid package we had offered.

Fearing I might have made a grave error in expressing my feelings, I used another phone in the Oval Office to call Secretary of State George Shultz while Reagan was on the phone with the White House switchboard.

"George, I think I really screwed up," I said. "I let this Beirut thing get to me too much and I told the president my concerns. He's calling Begin to get him to stop the shelling."

There was a brief pause. "Thank God," Shultz said. "I'm on my way." When the call came through from Israel, the president quickly picked up the phone. He told the prime minister, in very frank terms, that the shelling had to stop and that Israel was in danger of losing the moral support of the American people.

Reagan listened for a moment, before ending the conversation by saying, "It's gone too far. You must stop it."

Twenty minutes later, Shultz joined us, and Begin called back to say that he'd ordered Sharon to stop the bombings. There would be no more planes over Beirut tonight.

Reagan hung up the phone and stood up. "I didn't know I had that kind of power," he said with a wink and a smile.

If anything, I am most proud of that moment. Whatever my judgment was worth to him, I felt I had used it wisely to persuade the president to follow his own humane instincts. I believe that every president needs someone who has the ability and the relationship to talk frankly, even if it means telling the president that he's wrong.

For a time, a very brief time, Beirut was quiet.

One of my last jobs in the White House was to arrange a state visit to Germany. It was 1985, and Reagan had accepted an invitation from Chancellor Helmut Kohl to participate in World War II remembrance ceremonies surrounding the fortieth anniversary of the end of the European war. Both leaders rejected the idea of doing an event at a concentration camp. I knew Reagan's emotions would betray him in such a setting, and both men were anxious to project a forward-looking image of hope and reconciliation. Reagan simply could not perform in certain environments. During a campaign stop at a veteran's home in 1980, he was unable to complete his prepared text before losing his composure. He saw too many ghosts in the eyes of the ailing men in front of him.

I was charged with handling the advance work and planning for the entire trip. Along with our diplomatic team in Bonn, we selected what I thought was the perfect site to lay a wreath to remember the German war dead. The cemetery I reviewed was ideal, beautifully coated with a long winter's snowfall. It was called Bitburg.

Little did I know at the time, but lurking beneath the snow were the bodies of several dozen Nazis, hidden under unremarkable gravestones. Forget that many of these graves held eighteen-to-twenty-year-olds. The stones were clearly marked SS Waffen, all the proof needed that these dead soldiers had been responsible for unspeakably ghoulish acts during the war. As soon as the word was out to the worldwide Jewish community, all hell broke loose. In Europe, America, and the Soviet Union, the outcry was immediate and bitter. Wounds that never quite healed were pried wide open again. Veterans marched in angry protest and sent their medals to the White House in embarrassment. Jewish groups demonstrated, demanded—and then begged—

the president to reconsider. His initial reaction was a firm no.

I don't think I saw Reagan under more intense pressure than during the Bitburg crisis. My life became a living hell, so I can only imagine what Ronald Reagan was going through. Nearly the entire senior staff tried to get him to reverse his decision. Although he listened to his critics intently, he would not abandon his friend Helmut Kohl, who had called Reagan and asked him not to give in. Reagan didn't think the Kohl government could survive a presidential snub. What's more, he liked Kohl, a big bear of a German most people took to immediately. From the very first, they had been Ron and Helmut to each other. Reagan had also dealt with Kohl's predecessor, the pedantic Helmut Schmidt, and found the open and friendly Kohl a breath of fresh air.

I considered resigning, as if that would help, and I had my first serious rift with Nancy. Up to that point, I couldn't recall being on the opposite side of an issue from her, but now she was convinced that I had literally ruined her husband's presidency, and perhaps the rest of his life. I couldn't blame her either, since I secretly feared she might be right. It was by far the worst public relations crisis that happened to Reagan since he was elected, and it happened on my watch. Nancy phoned me in my office from the residence, and we had a very painful, emotional confrontation. I was already a wreck because I had let Ronald Reagan down. Disappointing my friend Nancy added to the devastation I felt. I let her finish her tirade and said, "Nancy, it's done. If going into a panic would help, then I'd panic. But I am trying my damnedest to make it right."

I somberly placed the phone back in the cradle. We would not speak again for some time. I know the president had an equally terse conversation with her, but she, too, was unable to make him budge.

Nearly every other senior aide from the California crowd joined Nancy in her crusade to stop Reagan from going to Bitburg. We were joined in our dissent by more than half of the

U.S. Senate. By coincidence, Elie Wiesel, the larger-than-life Holocaust survivor, was in the White House to receive the congressional Medal of Freedom. He lectured Reagan in front of an entire nation, passionately telling the president as a friend that his place was with him, not the Nazis. Incredibly, Reagan's resolve stiffened after the Wiesel incident. "The final word has been spoken as far as I'm concerned," he said afterward.

On my last trip to Bonn before the presidential visit, I decided to try one more time to get our German hosts to change the agenda, deleting the Bitburg event altogether. I was hopeful Kohl would acquiesce after hearing firsthand how many daggers Reagan was taking on his behalf. As I was getting ready to depart from the White House, my secretary buzzed me to say that a Matt Ridgway was on the line, and needed to speak to me urgently. I told her that I didn't know any Ridgway and besides I was just leaving to go to Germany.

She buzzed back. "It's *General* Ridgway." My God, I thought, is he still alive? Matt Ridgway was one of the country's greatest fighting men, serving with distinction in World War II and later taking over in Korea. He'd succeeded Eisenhower as supreme commander of NATO and served as U.S. Army chief of staff.

"Mr. Deaver, this is Matt Ridgway."

"Yes, sir," I said.

"I am a soldier, and I have never done anything political in my life," he went on. "But it appears to me that my commander in chief is in trouble and I would like to help."

I listened to him anxiously as he said, "I would like to lay that wreath in Bitburg for him. I am the last living four-star general who was involved in the European theater."

I put him on hold and walked into the Oval Office. I said, "Mr. President, you are not going to believe this, but Matthew Ridgway is on the phone and has offered to lay the wreath at Bitburg for you so you won't have to do it now."

I thought the ideal solution had fallen into our laps. One of the most bitter complaints had been directed at the prospect of the president of the United States appearing to honor the soil that contained the remains of the Nazi SS. Ridgway, in his nineties at the time, had led the U.S. 82nd Airborne Division in action against these guys and would defuse the issue. The glow on my face faded when I noticed that Reagan didn't share my glee. He picked up the phone.

"General, this is Ron Reagan. Mike just told me what you want to do and I can't let you. But I have a better idea. We'll lay that wreath together." Reagan seemed to take comfort in knowing Ridgway would be at his side, but I still wanted to get the president out of the entire affair.

This was my thinking as I headed to Andrews Air Force Base to board a government plane to Bonn later that night. Just as my car crossed the base gates, the phone in the car lit up. Happily, it wasn't the red phone—the secure line that only the president, myself, Ed Meese, Jim Baker, the national security adviser, and the military attaché who carried the nuclear codes had access to. This phone was just like any built-in, except that when you picked

it up, the signal corps or the White House operator was waiting for you. This time it was the
latter. The president wants to talk to you, his
personal operator told me.

"All right," I said. "Put him through."

"No, no. He wants to see you."

I asked her if the president knew I was on
my way to Andrews Air Force Base to fly to
Germany. Yes, she said, but he still wanted to
see me. It was almost midnight, and I was
forty-five minutes away from the White House.
Why couldn't I talk with him on the phone?
When the car finally got me back to the White
House, I went up to the second-floor residence
and found Reagan waiting by the elevator for
me, nearly in the dark, dressed in his pajamas. I
was about to start talking, but he cut me off
with a brisk sweep of the hand and led me into
his study. He was silent until he had shut all
the doors—Nancy was asleep in their bedroom, right next door.

"Mike, I know where you're going," he said.
He paused for what seemed an hour. Bitburg
had worn him out. Right now, he didn't have a

friend in the world, and here he stood—sleepless—trying to stop the one man who was trying to make the pain disappear.

"I know what you are trying to do," he said, "and I know you think you're doing what's best for me. You're going to talk to Kohl, and you're going to try to get him to call this whole thing off. What happened in Europe forty years ago was horrible, but it was forty years ago," he continued, more animated. "If we can't put that war behind us now, if we can't reconcile and look forward, then we'll always think about the war, never the peace."

I didn't say anything.

"I hope you understand what I'm saying?"

I nodded.

"I'm sorry to delay your trip, but I want you to get on that plane and tell Kohl tomorrow that there will be no change to the program," he said. "Tell him I'll stand with him as I promised." His voice rose from the whisper it was when he first greeted me. His strength was infectious. "We need to start moving forward, and we need to start now." Although he didn't quite say it, this was one of the rare occasions where he actually ordered me to do something.

Reagan went to Bitburg and stood with Kohl. He took a lot of arrows from some of his closest friends and almost everybody else, but he went. Right or wrong, he was there, a different drummer to his heart's deep core, and I'll leave it to the historians to judge the words he uttered: "We are here because humanity refuses to accept that freedom or the spirit of man can ever be extinguished. We are here to commemorate that life triumphed over tragedy and the death of the Holocaust—overcame the suffering, the sickness, the testing, and, yes, the gassings. . . . Out of the ashes, hope, and from the pain, promise."

I noticed early on that Reagan was the kind of man who fought policies, not people, unless they resided east of the Berlin Wall. Partisan labels didn't work with him. Even if you were against him on policy, you could be his friend after hours. He admires men for what kind of people they are. Democrat Thomas P. "Tip"

O'Neill was one of his favorites. Another Democrat, Dan Rostenkowski, once told me that a handshake from Reagan "would last him an entire week." These were political giants with whom Reagan had serious policy differences, but he wouldn't let that interfere with getting to know a good man. With Reagan, it was cocktails at dusk, pistols at dawn.

The relationship between President Reagan and fellow Irishman Speaker O'Neill was one of my favorites. They admired each other as adversaries but would always know when to stop tossing the grenades. In my West Wing office one day, I took a call from my longtime secretary, Shirley Moore. "Tip O'Neill wants to talk to you," she said.

I was confused. I knew little about Congress and even less about what the second most powerful man in Washington would want with me. I told him as much when I picked up the phone.

"Mr. Speaker, I think the guy you want to talk to is Ken Duberstein—he's our liaison with you guys," I advised, urging him to speak to our chief congressional affairs man.

❧

"Look," said the Speaker in a dismissive laughing tone. "I know Ken, but this is different. I hear you're the guy who can make things happen."

I listened as he told me that there was a U.S. Navy sub being christened the next day. It was to be called the USS *Corpus Christi*.

"Mike, do you know that Corpus Christi means body of Christ?" he asked.

I didn't answer, letting my silence serve as proxy for my inability to defend the name.

"That's the dumbest goddamn thing I ever heard," he volunteered.

He went on to outline his objections to the name, how it offended not only his Catholic sensibilities but also common sense. I couldn't argue with him. I did a little research and filled Reagan in on the problem. He put a call through to Secretary of Defense Weinberger on the spot.

"Cap, Tip's upset about this *Corpus Christi* thing, and I'm having a hard time finding fault with his logic," Reagan said. "Our intent was to name this sub after the town, right? So why can't we just call it the *City of Corpus Christi*?"

There were a few more pleasantries exchanged, the president hung up. "Cap says it's no problem," he announced proudly.

I called the Speaker back, and he took the call immediately. "See, I said you were the guy who could get things done." Tip was one of the grandest men I ever met in politics. Like Reagan, he was a partisan who also cherished civility.

Reagan invited O'Neill to a lunch in the residence on the occasion of the Speaker's seventieth birthday. Ken Duberstein and I listened to two guys cut from the same cloth. They were both Irish, but one chose to stay and fight in Boston politics while the other followed the stars to Hollywood. Other than that, they might have been brothers. At the end of lunch, Reagan ordered champagne and offered a toast.

"Tip, if I had a ticket to heaven and you didn't have one, too, I'd sell mine and go to hell with you."

The Speaker, known for his stoicism, teared up.

Today's partisans could learn a lot from their relationship. That is not to say that the Gipper got along with all of his political adversaries. Reagan's political instincts are sometimes underrated, but time and time again, he stunned me with his ability to make the right call at the right time.

One afternoon I fielded a call from Clark Clifford, a Washington graybeard who had met every president since Hoover. I had never met Clifford before, but like many people new to the capital city, I was aware of his reputation. The office walls of official Washington are pasted with photographs of starstruck Washingtonians who sought to befriend the Great Clifford. I guess I wanted to meet the legend, too.

My call with Clifford was short and to the point. He wanted to meet Reagan. In his courtly manner, he asked if he could come over following a meeting he was going to have with a White House functionary in the Old Executive Office Building. I looked at the relatively light schedule and told him that I'd get back to him later that morning.

I walked into the Oval Office and sat down across from Reagan. He was in a festive mood, but it soured somewhat when I brought up Clifford.

"Why would I want to meet with Clark Clifford?" Reagan asked, somewhat bemused.

I went into a mild defense of Clifford. He knew where all the bodies were buried in Washington, I argued; he would be a good guy to get to know. Reagan was still unconvinced. He usually deferred most scheduling matters to me, but for some reason, he resisted. I kept pushing.

Finally, he acquiesced. "Look, if you really want me to do this, fine," he said. "But I don't think any good will come of it."

Reagan would usually meet with anybody, Democrat or Republican, it didn't matter to him, but some sort of warning signal seemed to go off in his brain. He didn't want to do it but left it to me. When the day arrived, I walked

Clifford through the door. "Mr. President, this is Clark Clifford," I announced.

Reagan nodded and shook his hand. There was a short discussion, a typical courtesy call. I'd seen Reagan do these five-minute Oval Office walk-throughs hundreds of times. He was always gracious, and he was this time, too. But with Clifford the routine was subtly different. Uncharacteristically, Reagan stood by his desk and never offered his visitor a seat. I think he sensed that the old man had come to take his measure, and Reagan—a different drummer in ways both large and small—simply wasn't going to have anything to do with it. When Clifford left, Reagan gave me a half smile and shrugged his shoulders. I knew he wasn't upset with the meeting, but his expression asked: "What was that about?"

That evening at the house of another Washington institution, Pamela Harriman, Clifford was telling anyone who would listen, including the national media, what he thought of Ronald Wilson Reagan. Clifford said—

Governor-elect Reagan in a swearing-in practice session in Sacramento. His son, Ron Reagan, looks on, January 1, 1967.

(Courtesy Ronald Reagan Library)

The 1981 Inauguration swearing-in ceremony on the West side of the U.S. Capitol in Washington.

(Courtesy Ronald Reagan Library)

Reagan, seconds before the March 1981 assassination attempt on his life. I'm standing to his left. It is a terrifying moment and I'll never forget it.

(Courtesy Ronald Reagan Library)

Reagan, me, Dr. Daniel Ruge, and Ed Hickey aboard Air Force One, September 1981. Air Force One was always a good place to catch up.

(Courtesy Ronald Reagan Library)

Reagan and me at his beloved Rancho Del Cielo in California, November 1981.

(Courtesy Ronald Reagan Library)

Walking in the colonnade of the White House, October 1982.

(Courtesy the White House)

Reagan, me, and Jim Baker sharing a lighter moment in the Oval Office, February 1983.

(Courtesy Ronald Reagan Library)

My farewell announcement in the Rose Garden, May 1985. It was time to move on, and I would no longer work and live as closely with the two people who had figured so importantly in my life.

(Courtesy Ronald Reagan Library)

Reagan and Mikhail Gorbachev sign the INF Treaty, December 8, 1987. The walls were coming down.

Reagan departing the Capitol on his last day in office, January 20, 1989.

Our meeting in 1997.

(Courtesy the author)

famously—that he thought the president was "an amiable dunce."

I never talked to Clifford again. I had no interest in wondering why he'd say such a thing about a man he spent a total of five minutes with. In this instance, I was the dunce—probably not that amiable. Not surprisingly, Reagan never once gave me an I-told-you-so lecture about the meeting even though I had been the one who pressed him into it.

Reagan was intimidated by few, in awe of even fewer. One exception was Mother Teresa. Her visit to the White House was one of the few times I ever saw the Gipper a little anxious.

The lunch with Mother Teresa was on the schedule for several months, and Reagan asked me about it probably a half dozen times. He was fascinated with how she lived her life of service in the slums of Calcutta. He was also concerned that his Holy Visitor have a pleasant experience during her visit to the White House. We decided on a private lunch in the residence: just the Reagans, Mother Teresa, one of her fellow sisters, and me. It turned out to be one of the more unusual lunches I would ever have with Ronald Reagan.

After introductions and other pleasantries, we took our seats. Reagan looked at me, then at his lobster bisque. He then looked around again and picked up his spoon. As soon as the steaming bisque was in his mouth, Mother Teresa began praying very loudly, "Heavenly Father, pray for this man and this food we are about to eat. . . ." By the time we had hurriedly put our spoons down, the sister was crossing herself, completing her short, somber appeal to Christ. We sheepishly began again as Mother Teresa sat watching us in silence, her hands gripped together on her tiny lap.

"I've been looking forward to this meeting for a long time," Reagan said. "I've read about your work. Nancy and I admire you greatly."

Mother Teresa, though, was having nothing to do with small talk. "Mr. President Reagan, do you know that we stayed up for two straight nights praying for you after you were shot?" She pointed to the younger sister. "We prayed very hard for you to live."

Reagan was visibly touched and nodded solemnly. "Thank you for your kindness," he said, sincerely.

He was probably hoping to get off easy, but Mother Teresa responded in monotone, "You have been touched by the passion of the Cross; you must now dedicate your life to serving the poor." There was an awkward silence around the table. She couldn't have been more than five feet tall, hunched and all skin and bones, but Mother Teresa had the most piercing eyes I have ever seen—eyes like an eagle, and they looked straight into your soul. Now, as her voice raised slightly, her tone reflected her complete resolve:

"You must now dedicate your life to the poor."

"What can I do?" Reagan said quietly.

A small light went off in his head.

"Mother Teresa, you know we have a great deal of excess cheese and butter and other foods," he said triumphantly. "Would that help?"

"We take nothing from any government of any kind from any country," she said, cutting him off.

Another warrior against Big Government, I thought. Now finally Reagan could find some common ground. I was waiting for him to segue into the evils of federal largess, but he couldn't resist asking the obvious question.

"Where do you get all your food?" he asked.

She leaned her tiny frame closer to Reagan and pointed to his untouched boiled potato. "Mr. President Reagan, do you know how many people I could feed with that one little potato?" she asked rhetorically.

Reagan wanted to know and asked how many, as if more than one person could find use for this fist-sized offering.

"Ten," she said immediately. "Every night we go to the airport and sift through the garbage from the first-class cabins. We can feed five hundred people from the leftovers from one full first-class flight."

It was one of the few times I ever saw Reagan at a loss for words. He spent the rest of the lunch asking questions about her order and her ministry—what they did and how they served the poor. She left a sober-minded Reagan with the admonishment never to abandon his commitment to the sanctity of life and the unborn.

The Reagans had stopped going to church on a regular basis when they arrived in Washington: he could never get around the fact that his presence required normal worshipers to pass through metal detectors. But the president frequently attended private services at the White House and at Camp David, and those of us who had the privilege of traveling with him on Sundays got to know that he was a great hymn singer with a, well, "challenged" voice. Unlike Carter, Reagan was never one to wear his faith on his sleeve, but he felt it deep down.

In November of 1985, Reagan was preparing for another historic meeting, this time with the new Soviet leader, Mikhail Gorbachev. Reagan had longed to negotiate face-to-face with a Soviet counterpart, but as he said, "they keep dying on me." Gorbachev shared Reagan's zeal to sit across the table, and the first superpower summit in seven years was set for Geneva toward the end of 1985.

In some ways, the summit stakes were never higher. Tensions between the Soviets and the United States had been accelerating because Reagan was treating them as the adversary they truly were. He knew the Soviet system was morally bankrupt and said as much at every opportunity. Still, in Reagan's mind, this Gorbachev fellow offered a fresh opportunity to at least talk about reducing the thousands of nuclear warheads we had aimed at each other.

I had left the White House a few months earlier, but I kept in close touch with the Reagans. I also had a small role in helping to plan the summit, working with White House Chief of Staff Donald Regan and with George Shultz. A few nights before Reagan was to leave for Geneva, I called him to ask if I could stop by the White House for just a few moments. For some reason, he was less than enthusiastic about seeing me, or anybody, that night.

"Look, Mr. President, I'll only take five minutes," I offered.

He acquiesced and I arrived at the residence at about 6:30 P.M. Usually, when I met Reagan there, I'd get off the elevator and have to look around for him. But on this cold night, he stood about three or four feet from the elevator with a stack of briefing books tucked under his arm, as if to say, "See, I told you I was busy." Even more surprising, he didn't offer me a seat. I think it was the most preoccupied I had ever seen him.

"Mr. President, can we sit for just a moment?" I asked.

Reluctantly, he said, "Okay, but just for a few minutes."

I wanted to tell him something that was on my mind for the last couple of days. I knew that Reagan was working with a new group of top staffers. Most of the Californians had moved on to different lives, and I wanted to see him before he left and offer my unsolicited thoughts on dealing with Gorbachev.

"Mr. President," I began, "I've learned a

great deal from you. You've taught me that there is a purpose for each of us, and I think this meeting might be your reason, your purpose."

I was trying to appeal to Reagan's sense of destiny, making the point that this was his moment to start a new chapter in Soviet-U.S. relations. I firmly believed that Reagan's life might have been spared for this exact reason. If so, only he could take the next step toward peace. This got his attention. He nodded slowly, but his eyes remained focused on me. "I think that you need to get your team out of the room and get to know this guy one on one," I continued.

"I think that may be a good idea," he said, starting to smile.

It turned out Reagan had already thought of meeting with Gorbachev alone, arranging for a private session in a pool house on the estate where the meeting was taking place. I had simply helped to refocus and confirm the correctness of his instincts.

The summit's results have been well documented, and Reagan's transformation from cold warrior to peacemaker changed the world. About a week or so after his triumphant return to Washington, he called me and asked if I would stop by the residence for a debriefing. His festive tone was infectious, and I was at the White House inside an hour.

I congratulated him on his masterful handling of Gorbachev. What had he learned about the Soviet leader during their brief time together, I asked, especially alone in the pool house?

"He believes," came the response, almost a whisper.

I looked at him completely befuddled. By definition, communist regimes were utterly contemptible of anything religious. "Are you saying the general secretary of the Soviet Union believes in God?"

"I don't know, Mike, but I honestly think he believes in a higher power," Reagan said.

The Cold War had been raging for decades. During that time, countless Kremlinologists at work in the State Department and the National Security Council had ginned out reams of top-secret data analyzing every Soviet leader. Some of it ranged from useless to such absurdities as his preference for coffee or tea.

Sure, Reagan read these documents to learn about Gorbachev, but he was looking for something else, and he found it. Reagan knew, without benefit of a conversation, that Gorbachev was a different kind of Soviet leader, not another of the gray, soulless apparatchiks who kept dying on him. With Gorbachev, Reagan could do business, and business they would do, eliminating entire classes of nuclear weapons and paving the way for the literal collapse of arguably the greatest enemy we have ever faced.

Ronald Reagan saw a saint in nearly everybody, bad in almost nobody, especially his staff. This probably isn't the best quality to have in a president, but he survived because he had men like Ed Meese and Jim Baker at his side. They put Reagan first, not their own personal agendas. Eventually, though, the Big Three—if I can count myself among them—would burn out, with dire consequences.

By 1985, the Reagan presidency was being served by strangers, newcomers who brought it to a standstill through bad staffing decisions. Nowhere was the damage greater than with the Iran Contra scandal, a body blow to the administration and to the man himself. The scandal involved selling arms to the Iranians in exchange for Americans being held hostage in the Middle East. These monies were then diverted to the Contra rebels fighting in Central America, in violation of federal law.

I was out of the White House at the time, but thought I should make the case to the president that Chief of Staff Donald Regan needed to go. Although Regan was not implicated directly, the scandal happened under his watch. I agreed with Reagan's friends that change was required if the country was going to put the crisis behind. Reagan, as usual in such matters, was immovable. He resisted his friends in Congress who called for Regan's scalp. He resisted even Nancy, and believed that the scandal would blow over. Reagan said repeatedly, to them and to anybody who would listen, that

Regan did not deserve to be "thrown to the wolves."

Toward the end of 1987, I enlisted my old comrade Stu Spencer to give Reagan an outsider's perspective. We met with him in the residence. I knew what buttons to push, but I'd have to tread lightly.

"Mr. President, this is difficult, but you're not the first president to face tough decisions," I said, again outlining my rationale for a clean slate.

Reagan became livid. "I'll be damned if I'll throw somebody else out to save my own ass," he shouted as he threw his favorite fountain pen to the ground.

Neither Stu nor I said anything for a moment.

I piped up. "It's not your ass I'm talking about, *Ron*," I said, "it's the country's ass."

I had just rebuked the president of the United States. Worse, I thought at the time, I just called him "Ron." What in God's name was I thinking? Eventually, I would find out that the presumed intimacy hadn't bothered the president that much. He didn't stand on ceremony with his friends, even when he was the first among them. At the time, though, I was just determined to charge forward with what I had to say. I reminded Reagan that he had taken a sacred oath on the steps of the Capitol. He needed to think about the country, I said.

At this he became quiet, almost whispering, "You know, I've always thought about the country first."

As Stu and I limped out of the residence, we both realized that Reagan was too angry to give in. He needed to make the decision in his own way. He would not be rushed by us or anybody else. Eventually, he pulled the trigger, or let it be pulled for him. Either way, Regan would leave the White House.

Throughout the entire Iran Contra affair, Reagan believed what he did was right and that he was telling the truth to the American people. Of this I have no doubt. I never heard Ronald Reagan tell a lie—not once. I think it would have been impossible for him to do so.

* * *

I've been inundated with requests to compare
Reagan with other recent presidents, especially
Bill Clinton. While I do want to keep the focus
on Reagan, a brief comparison of the two
men's political skills merits discussion. Regarding their differences, I have found that,
like most politicians of our time, Clinton is a
work in progress. He didn't come to office with
a set core of beliefs. Reagan, on the other hand,
moved into the White House having said essentially the same thing for two and a half
decades. He spoke incessantly of America's
grand potential, that government needs to be
tolerated and managed by the people, not the
other way around, and that freedom super-
seded almost anything. During the campaigns
people used to beg me to let Reagan speak
on another topic. I would often yield, allowing
these energetic and talented speechwriters
to develop a piece on, say, federalism. They
would do outstanding work and Reagan would
use the speech, but as he spoke, he would always find a way to get back to his favorite

themes. This old dog would not be learning any new tricks. Clinton, as we all know from his State of the Union marathon sessions, is totally at ease in introducing new issues at the drop of a hat.

They do have things in common. Both are tireless campaigners. Both genuinely love people. And Clinton, like Reagan, feeds off a live audience. Put both men in front of a joint session of Congress or at the podium of their respective conventions, and watch how they electrify the room.

In smaller settings, there is a marked contrast. Unlike Clinton, Reagan never "worked" a room. He'd stay with the person he was talking to and throw all his attention that way. He would never peek over his conversation partner's shoulder. Big or small, important or just an Average Joe or Jane, this person became the subject of his interest, and the rest of the room was a blur.

They were different, too, on those occasions when the president needs to address the nation from the Oval Office, often in times of crisis.

There is no audience, just a still camera sitting a few feet from the desk. Clinton is good, but I don't think he likes to be alone with the lens—he's only given one Oval Office address. Clinton can't touch the Gipper in this respect—few people can. Reagan had the Hollywood background and the radio stint, but more important, he instinctively understood that there were real people on the other side of that little lens. To make sure he'd remember this, he would literally identify three of four real people that he knew would be watching the speech and pretend he was talking directly to them.

Finally, there are myriad examples of Reagan's inner strength that the Clinton generation simply won't understand. Sure Reagan would allow his political people to take polls from time to time, but I don't think the phrase "overnight tracking" ever crossed his lips. If he were in the political arena today, he'd be amazed to learn that a presidential candidate might actually focus-group-test single words and phrases like "risky" and "scheme" to see how well they reverberated with the public. Needless to say, nobody would have had the nerve to ask Ronald Reagan, "Boxers or briefs?"

I firmly believe that God watches out for this world in which we live. Maybe He's a little biased toward the United States, because the right people just seem to come along when they're needed most.

In 1980, Americans were under siege, afflicted by a "malaise" that was tightening its grip in the form of long gas lines, insufferable inflation, and a dwindling national pride. Twenty- and thirtysomethings can't know what waited for Reagan when he stepped into office in January 1981. Hostages in Iran, Soviets on the move in Central America and the Middle East, an economy in the tank, the U.S. military flat on its back, relentless foreign competition bruising our greatest companies. The American flag was an object of scorn overseas. We were on the verge of becoming a laughingstock.

Americans rightly were beginning to doubt themselves, but a man came along who knew the true score. He knew that better times were ahead because "after all, we're Americans." The nation cried out, and a guy named Ron answered the call. He resuscitated not only the diminished office of the president, but the national pride that had lain dormant since the early 1960s. Clearly, Reagan was the right man at the right time, and history will bear his name triumphantly.

I firmly believe that Reagan's eight years in office changed the way people look at their commander in chief, their head of state. He made it okay to look at the White House or Old Glory and get goose bumps again.

Reagan once said that he couldn't imagine life without Nancy. I guess I can't image what my life would have been without him. Despite my ups and downs with the man, I would not trade one minute of my time with him. Just after he won the White House in the 1980 election, I took a call from H. R. "Bob" Haldeman,

Nixon's former chief of staff, who invited me to lunch.

"What job are you going to take?" he asked with great interest. I outlined my reservations about going to Washington, telling him about my young family and my love for California.

He had a look of disbelief on his face. I thought if anybody was anti-Washington, it'd be this guy. I couldn't have been more wrong. "Mike, there have only been forty presidents, and each one of them had one special guy on their staff. You have a chance to be that guy."

I needed to know if it was worth it. After Bob's Watergate humiliation and the pain of imprisonment, I asked him, "Would you do it again?"

"In a heartbeat," he said instantly.

I never got a chance to thank Haldeman, but he changed my mind. I got to be Ronald Reagan's guy for five years, serving at the side of one of the greatest presidents of the twentieth century, the American Century. It's something that I'll never forget. I'll also never forget how incredibly, how perilously close this whole wild ride through Washington had come to being cut off in its infancy. For the last two decades, the sound of John Hinckley's gun has never really been out of my ears.

CHAPTER FOUR

A Bad Day in March

March 30, 1981. It's funny, fate put me at the president's side that day, but I never think consciously about it unless asked. Even now, when retelling the story, my emotions usually get the best of me. There are a lot of reasons for the pain—the trauma of the day; the fact I was in such close range to the assassin; the many good men who were wounded, one grievously; and that I almost lost my friend. Ronald Reagan had been in office less than ten weeks when he was shot by John Hinckley Jr., barely time to get to know his new job, much less begin to exercise the powers of his office. Had he been killed—and I still shudder at how close that came to happening—his presidency would have been another of those tragic chapters that sometimes mar our great national story, this one notable by its astounding brevity. Instead, Reagan's shooting and his recovery became a study in hope, renewal, and personal courage. I consider it one of the great blessings of my life that I was on hand to watch this astounding man in what could have been his darkest hours.

Even though the president's recovery time was that of a young man, profound debate among reasonable people continues about Reagan's effectiveness in office after the attempt on his life. The charge was put back in play in late 1999 when Edmund Morris's book *Dutch* hit the bookstores. In appearance after appearance, Morris, a man of great intellect, contended that Reagan was essentially given cold blood during a transfusion immediately after the March 30 shooting. Because of the need for immediate surgery, Morris tells us, there was no time to adequately warm the blood that Reagan would be receiving. The cold blood, he says, was a dramatic insult to the president's body, similar to a prolonged beating. Reagan, Morris alleges, never fully recovered from the experience.

Although the cold-blood theory occupies only a small space in a hefty tome, the words did their damage. They have once again awakened Reagan's critics from their slumber. Like the endless search for a single gray hair on the presidential head, Morris's ruminations on transfusions have helped stoke the fires of those who would discredit the Reagan presidency.

The charge—and those that rose from other quarters—is a serious one, an indictment of the efficacy and legitimacy of the Reagan years. For history's sake it merits the strongest possible rebuke. Since Morris didn't dwell on it in his book, I can only guess that his publicists used it as a means to stir debate among Reagan loyalists, thereby driving up book sales.

The reader doesn't have to take my word. Dr. Joseph M. Giordano, the director of President Reagan's trauma team, and Dr. John E. Hutton

Jr., the president's physician, cleared the air in a *Washington Post* column on October 15, 1999. "Apparently Mr. Morris is concerned that the cold blood had a long-term effect on the president," they wrote. "Blood is not stored in freezers, as stated in Mr. Morris's notes, but usually kept in a refrigerator. By the time blood is administered through IV lines, the effect on the core body temperature would probably be small."

The doctors continued: "At no time since undergoing his lifesaving left thoracotomy in 1981 and his remarkable recovery has President Reagan shown any abnormality in his blood profile. This would include the results from his annual physical examinations and four subsequent hospitalizations for surgical procedures. After each procedure, his recovery was extraordinarily rapid and never with an untoward complication. All of his physicians have agreed that President Reagan was among the most resilient of patients they have ever encountered."

I take responsibility for introducing Morris to the Reagans. I was drawn to him by the eloquent prose in his Pulitzer Prize–winning treatment of Theodore Roosevelt, and I had heard his praises sung by some of Washington's most astute people, including Selwa "Lucky" Roosevelt. The Reagans met Edmund and Sylvia Morris—she has a Pulitzer Prize as well—at a small dinner party hosted by Senator Mark Hatfield and his wife, Antoinette. They were charmed by his broad knowledge of many subjects and his sense of humor. Like me, they greatly admired his work on Teddy Roosevelt. All of us had hoped that both the public and future historians would be well served by granting Morris rare access to the president. Although some parts of his book, especially the later years, make for wonderful reading, it was clearly an opportunity lost.

I began to get bad vibes in the mid-1990s when I was sharing a dais with Morris at a media roundtable at the National Press Club during a discussion of the upcoming PBS *American Experience* program on Reagan. Morris laid out an outlandish, unsubstantiated claim that Reagan's Alzheimer's was brought on by the assassination attempt and that Ronald Reagan went downhill thereafter and was never the same.

I was stunned. I had never heard a qualified physician, let alone a writer, make such a claim. I immediately rebuffed Morris, making it clear during the program that there was not one iota of evidence to support the claim. I never heard the accusation again, either in print or on television, but it is clear to me now that Morris was determined to make sure history showed that Reagan was a changed man from the wounds he suffered.

When the Alzheimer's trial balloon sank, Morris moved on to the contention that the Gipper was never the same after receiving the frosty blood of strangers. Both are patently wrong, but the blood theory is the more dangerous because it does more damage to history and to Reagan's legacy.

What's the real story? Well, I don't pretend to any more medical knowledge than Edmund Morris has. Nor, certainly, is my prose any match for his. But what I lack in expertise and writing grace I make up for in proximity. I was there for the attempt on the president's life and with him nearly every day for the four and a half years thereafter. Although it was nearly twenty years ago, I remember it like it was yesterday. I have no great desire to revisit those moments—I've done enough of that—but the contention that Reagan and his presidency were somehow miniaturized by a would-be assassin's bullet leaves me no choice. I won't deny that it was a difficult period, but Reagan recovered and his presidency was one of the most successful, and important, of the century.

March 30 started off about as routinely as possible in the Oval Office. The president showed up in the West Wing at his standard time, 8:45 A.M., to meet with his top staff for his morning briefing. Several weeks earlier, newly arrived in the capital, I had locked him into giving a stock speech to a couple of thousand of generally sympathetic labor representatives who had now descended upon Washington. The rank and file of organized labor made up large chunks of the Reagan coalition—the famous Reagan Democrats—and the speech was a way to thank them in person. We also needed their help in Congress for Reagan's economic recovery program. Besides, it was a rare light day on the president's schedule, and the Washington Hilton was just a short trip up Connecticut Avenue.

A little after the lunch hour, Labor Secretary Ray Donovan arrived at the White House to make the short trip with the president to the Hilton. As usual, those making the journey with the president would do so in a caravan of thirteen to fifteen vehicles. At least half the caravan was devoted to the traveling press corps and support for the Secret Service. For national security reasons, only four vehicles mattered: a lead car that contained local police and the Secret Service; the second car with the president (on this day, I was in this car); the third, usually a large Suburban, carrying an army of Secret Service agents; and a fourth staff car containing the president's personal physician and a military aide who kept the nuclear codes with him at all times. These four vehicles—the "package" in Secret Service parlance—were never to be separated. Any attempt to break up the package would be viewed as an immediate threat to our national security. During the height of the Cold War, the Secret Service had to be prepared for anything. On this day, they would be tested.

The president's car pulled up and was parked at the curb, several yards from the hotel's VIP entrance. Reporters were already positioned behind a velvet rope in hopes of firing off questions to Reagan after the quick speech.

The appearance itself was standard issue. The president waited in a holding room while the head table got organized, came out and gave his speech, and waited briefly again in the holding room, while the traveling press piled into their vans and the motorcade got ready to roll. The whole affair was over in a relative wink.

We started out of the Hilton at 2:25 P.M. We had been away from the Oval Office for only about an hour. I was walking out in front of the president when I remembered the gaggle of stalking reporters roped off by the door. I was just completing a joke I'd gotten earlier in the week from Vice President Bush when the first question was shouted out by Mike Putzel, of the Associated Press. Without thinking, I did what I always did: I grabbed Press Secretary Jim Brady's arm and steered him toward the rope line harnessing the media. I followed Jim out of the hotel, next turning left toward the car, with two Secret Service men and the president behind me.

I walked around Brady, who was dutifully ignoring the reporters, on my way to the far side of the car. I had reached the right rear fender when a member of the press corps bellowed out a question I can't remember. Naturally, Reagan did what he always did. He cracked the patented smile while raising his left arm in a friendly wave. Only Reagan could acknowledge the voice and dismiss the question at the same time without looking bad.

Then I heard the first pop. Later, everyone would agree that it sounded like little firecrackers going off. The smell of sulfur told me differently. I quickly ducked, then went all the way down and froze as more shots were fired. As it turned out, Hinckley was shooting literally over my right shoulder. Most of what happened next comes back to me in slow motion, maybe because I have seen the video footage more times than I care to count. For a time, even as recently as the mid-1990s, it seemed like I couldn't escape it. It stalked me like a bad dream, popping up in such unexpected places as in the movie *Forrest Gump*.

The picture of Reagan's face as it was caught on that videotape is unforgettable. There he was standing with his arm in the air as the bullets began flying. He had no idea he'd been hit—nobody did—but the smile was clearly gone, I can assure you of that much. His face was frozen in time, the smile replaced with a look of absolute helplessness.

When the shots ended an eternity later, I raced to the left rear of the president's limo and tugged at the door, but at the first sign of danger, Reagan's driver had thrown the automatic locking mechanism. Only the door on the rear right was open, waiting for POTUS—the President of the United States in Secret Service shorthand.

On the other side of the limo, veteran security chief Jerry Parr had already gone into action, diving into the president and breaking his frozen expression. Parr's job was to take the assassin's next shot, to literally become a human shield, and that's just what he had done. His momentum had taken him and Reagan onto the floor of the waiting limo via the open door on the right side. As it sped away from the hotel, Parr remained on top of the seventy-year-old man out of absolute fear of what might happen next.

I ran to another waiting car—the control car—and jumped in. As we gave chase to the presidential car, images of the people we had left behind came flashing at me: Secret Service agent Timothy McCarthy had been lying on his side, holding his stomach, with blood on his shirt when I last saw him. A D.C. patrolman I would later learn was named Thomas Delahanty was also down. Worst of all was Jim

Brady: I can still see Jim lying on his stomach with his face turned toward the limousines. His eyes were staring straight ahead as a pool of blood spread beneath his head.

I had no time to process the sight I had just witnessed; it was still only random, horrible images that existed outside of reason or logic. We were moving so fast down Connecticut Avenue that the landscape blurred and my thoughts discombobulated. Within minutes, I thought, surely we'd be back in the safety of the White House compound where I could see firsthand how badly Reagan had been hurt after being leveled by the flying Parr. From my vantage point, I could see the president sitting up in his car, but I couldn't contact him. The Secret Service was employing radio silence. We could have been in the middle of a conspiracy and were forced to take every precaution.

When we pulled into the emergency lane at the George Washington University Hospital instead of the White House, radio silence was over and the Secret Service apparatus was in full gear. Word was that "Rawhide"—Reagan—was not shot. Indeed, inside the limo the president was acting like anything but a man whose life was threatened. As the motorcade first roared down Connecticut Avenue to return to the White House, he had let loose at Jerry Parr, complaining that the Secret Service agent had thrown him so hard to the floor of the car that he'd probably broken a rib. When Reagan paused for a second to cough up a pinkish froth of blood, Jerry immediately redirected the limo to the hospital. This was a man with blood in his lungs; whatever had punctured them—bullet or busted rib—the president needed help fast. Parr's quick call probably saved Reagan's life and changed the course of history. To this day, whenever I see Jerry, we share an unspoken moment of painful recollection, reliving the day by simply looking in each other's eyes.

Our control car pulled up almost immediately after the president's car. I saw Reagan get out of the car unaided, look to both sides, and give a tug to his pants. I think he actually buttoned his suit jacket. So far, so good. He walked toward the emergency room doors unassisted, with a pair of agents at his sides, but as soon as he was through the doors, out of public view, the strength in his legs abandoned him. Thankfully, the agents were there to catch him and help him to a private room. I have always been grateful that the agents let Reagan walk in on his own. No one, not even a would-be assassin, was going to bring Ronald Reagan down.

The president of the United States had just walked into a totally unsecured emergency room, and none of his senior aides had any idea how bad things really were. I quickly got on the phone with Chief of Staff Jim Baker and Ed Meese, counselor to the president and, like me, a Reagan man from the Sacramento days. I told them all I knew at the time, that Jim Brady was in horrendous shape and the president had probably broken a rib and perhaps pierced a lung. The doctors were with him as we spoke. As I relayed the events that had unfolded on Connecticut Avenue only minutes earlier, two ambulances arrived at the same time. They wheeled Jim Brady out of the first. Agent Mc-

Carthy, who took one to the chest, was in the second. The D.C. police officer had been routed to another hospital. I was watching them unload my fallen colleagues when an intern approached me and asked without emotion, "Do you know the patient that was just brought into the ER?"

I couldn't believe the question. "Yes, I do," I responded incredulously after figuring out that he was completely serious.

Frustrated with me, he said, "Would you give me his name, please?"

I said in monotone, "It's Reagan." I even spelled it out for him. "R-E-A-G-A-N."

I waited for an earth-shattering reaction, but all I got was, "First name?"

When he asked for the address, I lost my cool.

I think the tree finally fell on him and he said, "You mean . . . ?"

While I was jousting with the clueless intern, Baker and Meese were trying to secure things at 1600 Pennsylvania Avenue so they could come to the hospital. I asked them first to call

Nancy Reagan and persuade her not to come. After all, if it was just a badly bruised rib, we could be out of there in an hour or two. Had I known what was really going on behind the door, I would of course have told Nancy to come, but at the time I sincerely believed that President Reagan was not in any real danger. (Somewhere in the confusion of those first hours at the hospital, Dr. Daniel Ruge, the president's physician, would suggest to me that Reagan might have suffered a heart attack. Almost anything seemed possible at the time, but I told the doctor that I had trouble believing it—Ronald Reagan was strong as an ox.) Regardless, my plan to keep Nancy at bay was already moot. She was being briefed by the Secret Service while Jim, Ed, and I talked. "Rainbow"—her Secret Service moniker—would soon be on her way.

After I hung up the phone, I ducked into the room where the president had been taken. The sight was a chilling one. My friend, my president, was stripped to the skin, and one of the doctors was holding his coat up to the light, evaluating the tiny bullet hole under the left sleeve. Reagan had lost all his color; his eyes open and moist. He was staring blankly at the ceiling. Shaken, I leaned on the doorjamb for support.

Only fifteen minutes had passed since I put the first call through to Baker. I called him again now with sobering information. "It looks like the president's been nicked," I said. I was just beginning to tell Baker that the news was far worse on Jim Brady when Brady was wheeled by on a stretcher. The hall was tight enough that I had to squeeze in my stomach so the gurney could get by. When I looked down and saw Jim's savaged head, I nearly passed out. His head seemed wide open, the blood soaking the pillow beneath it. I managed to find my voice, even though it was cracked and strained, to tell Baker and Meese that "It doesn't look good for Jim." In a day of awful calamities, though, there was at least one good surprise. The hospital was holding its monthly meeting of department heads. When word reached the meeting of the disaster in the emergency ward, the chief brain surgeon and thoracic surgeon rushed down to take over.

Watching the news on a small television at the hospital, I realized that Brady and I actually look somewhat alike. In the confusion, I had not thought of calling my wife, Carolyn, and my two kids to tell them that I was okay. What if she saw the tape or heard from a friend that I was hit? I called home but got no answer. Finally, my secretary, Shirley Moore, was able to get through and assure her that I was unharmed. I never held Carolyn and my kids tighter than I did that night.

When Mrs. Reagan arrived at the hospital, she was still unaware that her husband had been shot. With great care, the presiding doctors told her that he had taken a bullet but was doing well. They admitted that they had yet to locate the bullet but were preparing to operate immediately. I stood back and carefully studied the look on Nancy's face. She stayed strong during the bullet revelation but was reduced to tears when they wheeled her husband to the operating room. His first words to her were: "Honey, I forgot to duck."

By now, Baker and Meese had arrived, and through the wall of green smocks the president spotted us. His first words were "Who's minding the store?" We forced out a collective nervous laugh. At that, the orderlies took him into surgery. I touched Nancy's arm and asked her to join me at the tiny chapel upstairs. We prayed together with the loved ones of the others who had fallen.

Afterward, we were escorted into one of the doctors' offices to wait and watch the news unfold on television. It was there we heard the first erroneous report of Jim Brady's death. Recalling the grizzly scene I had witnessed only a few hours earlier, I wasn't surprised at all by the report. I didn't see how he could possibly live. In the crush of events, some of the newsmen had made understandable errors, but ABC's Frank Reynolds would have none of it, berating on air any of his staff who passed him bum information.

Earlier, the police and the Secret Service had asserted control, securing the hospital and relocating the other patients to places unknown. Baker and Meese joined Nancy and me in the doctor's office, to wait for some word, any word, on the success or failure of the surgery. It was at this point that we debated the need to invoke the Twenty-fifth Amendment and pass presidential power to Vice President George Bush. Baker spoke with Dr. Ruge, who reported the president had lost lots of blood but was stable.

At the time of the shooting, the vice president was in Texas giving a speech to a cattlemen's convention. He was to fly from there to Austin to address the state legislature and was in the air when he received word that the president had been shot. The pilot made a beeline for Washington. We decided against invoking the Twenty-fifth Amendment at the time, but it made the three of us talk about the unthinkable. What if the president died?

Early in the evening, about five hours after the shots were fired, the president was brought to the recovery room under an orange blanket. I approached him as he slept, but he was barely recognizable. His skin was gray and drawn, his breathing labored. I began to feel that the doctors were putting on an act with their encouraging reports. I confided to Carolyn later that evening my fear that even if he got through this alive, he'd never be the same. We cried together. It would take time for me to understand and grasp the man's remarkable strength.

Of course, I said nothing of these feelings to Nancy. The thoracic surgeon, Dr. Benjamin Aaron, came out of the operating room and said to her, "It took me forty minutes to get through that chest. I have never seen a chest like that on a man his age." The compliment did little to comfort the nervous First Lady, but Dr. Aaron was articulating what I already knew and what the public would come to see so clearly in the days and weeks ahead: that Ronald Reagan was one tough customer. The remarkable acceptance of at least the first six years of his presidency and the astounding personal popularity Reagan was to enjoy began to take shape that day, I think—born out of his raw physical courage and the grace and aplomb he was to show under circumstances almost impossible to conceive. In his near-end, to paraphrase T. S. Eliot, was his beginning.

Reagan didn't leave the recovery room until a little after six the following morning. We walked in on him—Baker, Meese, and I—a half hour later. He had a tube in his throat and could not talk without discomfort. The nurses were still buzzing over the handwritten notes he foisted upon them from his pad of pink paper.

One read, "I'd like to do that scene again—starting at the hotel." After sleeping for a while, he dashed off a question: "I'm still alive, aren't I?" He followed that up with, "Winston Churchill said there is no more exhilarating feeling than being shot at without result." He was on a roll. "Send me to L.A. where I can see the air I'm breathing."

The tube was removed from his throat at three in the morning; finally able to talk, he wondered about the assassination attempt. He said he recalled hearing three or four rounds.

He asked if anybody else had been hurt, but nobody would give him a direct answer for fear of slowing his convalescence. He had been hit—it turned out—by a fragment of a bullet that had rebounded off the armored car door. The bullet pierced him below the armpit, traveled several inches down his left side, bounced off a rib, punctured his lung, and ended its brutal journey just next to the presidential heart.

The attempts by the networks to explain the medical details resulted in still more confusion. One network botched it, saying the president underwent open-heart surgery. He hadn't, but the doctors had opened his chest to find and remove the bullet.

When the three of us walked into his room, he was propped up in his bed, brushing his teeth. "I should have known I wasn't going to avoid a staff meeting," he said. When I told him that the White House was running like a well-oiled machine in his absence, he retorted, "What makes you think I'd be happy to hear *that*?" Our visit was more than a pilgrimage to an ailing friend. We had brought with us a dairy bill for his signature. There was no hurry to sign, but we were convinced the public needed to see that the Reagan presidency would not be sidetracked because of John Hinckley Jr.

Reagan grabbed the pen with zest but meekly signed the bill. Looking at the moist ink, I would have sworn it was a fake if I hadn't been there. Still, the small gesture showed that he was at least symbolically on the job.

Reagan had been shielded from newspapers and the cameras, but that afternoon it was decided that he needed to be informed of what had happened to the others who had been shot in the line of duty. The job fell to Dr. Ruge, who advised Reagan of the near-fatal injuries to Brady. Reagan was reduced to tears when he heard that a bullet was lodged in the brain of his press secretary. The assassin had been firing at nearly point-blank range with a .22-caliber handgun, a small gun but one designed for deadly accuracy.

"I began to pray for Jim and Tim," Reagan told me later, "but I realized if I was going to do what was right, I'd have to pray for that boy who shot us, too."

Jim Baker directed me to make sure that Jim Brady continued as Reagan's press secretary until he left office. This symbolic but warm move allowed Brady to retain the title and pay for the next seven years although he would never again be able to function like his old self. Jim also asked me to arrange a fund for those injured, a project I worked on with Democratic wise man Bob Strauss. It helped those who had fallen and remains in place today for anybody on a presidential staff that, God forbid, meets the same fate as Brady and Tim. Today, Brady remains confined to a wheelchair thanks to Hinckley. Meanwhile, the would-be assassin's lawyers are seeking twenty-four-hour furloughs from Saint Elizabeth's mental facility for their client. I'm no lawyer, but I had a firsthand look at the havoc Hinckley caused. Either he was sick in 1981 or guilty of attempted murder. Which is it?

The doctors had predicted that the president would need at least two weeks to recover sufficiently from his wounds to return to the White House. Dismissing them with a wave of his hand, he informed them that he was a fast healer. Despite Reagan's iron will, though, he remained in the hospital for twelve days. His room became a virtual White House annex. Nearby, a special communications apparatus was established and a desk brought in for his secretary, Helene Von Damm. We moved our morning staff meetings to the hospital's colorless cafeteria, working over cereal and doughnuts. Nobody paid any attention to us in our IBM-like blue business suits, probably thinking we were a couple of ambulance-chasing lawyers or doctors, when in reality we were helping to run the executive branch of the federal government.

The government hardly skipped a beat during the president's recovery, in large part because of the Reagan style. He's truly a big picture man who never enjoyed toiling in the details. He had guys like us to work on the small stuff, and he had a vice president built for this kind of potential crisis. The country owes a debt of gratitude to George Bush. He took over many of the president's day-to-day duties and handled the job masterfully, without once appearing to be a man longing for the job.

This wasn't some cold political calculation on Bush's part; it was simply the kind of vice president he was, and the kind of man he is. During the cabinet meetings he chaired, Bush sat in his usual chair, leaving the president's place at the table symbolically empty. He wouldn't even allow a photograph to be taken of himself that included the White House or any of its offices as a backdrop.

When Reagan was finally able to return to his rightful place in the Oval Office, the entire White House staff and dozens of administration officials and their families gathered on the South Lawn to greet him. Acknowledging them upon his triumphant entrance, he instinctively raised his left hand in an easy wave; it was the same wave I had seen at the Hilton. I winced at the irony, wondering if he made the connection, too. I doubt he did; that was already old news. Besides, there were better images to soak in, especially the striking rainbow that arched over the White House that day like a welcome-home gift from above.

We had persuaded him to limit his workday to a couple of hours until he was back to full speed. Reagan confined himself to the family quarters, with an occasional field trip to the Rose Garden for some sun. He once told me that he had had a recurring dream all his life that he would live in a house with very high ceilings and white walls, and the second-story residence at the White House fits that dream to a tee. It's a big barn of a place, with rooms opening off on either side of a giant hall. Nancy always kept the hall uncluttered from end to end, so you could feel its epic expanse, but she made the rooms warm and inviting by bringing in beautiful antiques and paintings that had been kept in storage and by filling them with flowers, especially orchids. If you had to recover from a near-fatal shooting, it wasn't a bad place to be, especially once the president was well enough to pad up to the third-floor solarium in his silk pajamas, bathrobe, and slippers. Enclosed in glass, the solarium was always a favorite spot for both Reagans for a late Sunday breakfast and a leisurely tour of the newspapers.

During this time, the president's schedule was limited to a pair of meetings in the morning: a session with Baker, Meese, and me, followed by a national security briefing. The meetings seemed to be the same, but the person leading them was in his jammies, not that this was a rare sight for those of us who worked closely with him. Reagan was never self-conscious. You could call on him most nights he didn't have an engagement and you'd find him wearing his nightclothes. The Reagans are the biggest pajama fans this side of Hugh Hefner. On nights when he had a major speech or press conference, Nancy would try to talk him into hitting the hay for an hour in the afternoon. Reagan said he'd heard that LBJ was a big proponent of afternoon napping. "Must have been something to it," he said, but he seldom did.

During the early stages of recovery, the president received just one outside visitor. The circumstances were unusual. On the morning of Good Friday, just after our staff meeting, he turned to me and said he'd like to see a man of the cloth. Nothing more was said. I was accustomed to such requests popping up without warning. I went to my office and called the Archdiocese of New York, the office of Cardinal Terence J. Cooke. Reagan had struck up a friendship with the cardinal, presumably based on their shared zeal for pro-life causes. I think their first meeting was at the annual Al Smith Dinner in New York City. The Democratic Smith was the first Catholic to run for president. Reagan was there and made some lasting friends.

I put the call in on Good Friday, maybe the busiest time of the year for a church leader. It was like calling an insurance man after a hurricane, I thought as I left word asking if the cardinal would consider coming down to minister to President Reagan. He was on the next shuttle flight to Washington.

I sent a car to meet him at the airport and bring him to the White House residence. Reagan then led him into the second-floor Yellow Room, where foreign diplomats often meet the president of the United States for the first time. They sat side by side on a yellow silk settee for nearly an hour while I waited outside, concerned that the president might tire or need something. As I walked in unnoticed, they were still in a deep, hushed conversation. I was privy to some of their closing words, including what the president promised the cardinal.

"I have decided that whatever time I may have left is left for Him," Reagan said. I waited awkwardly for a few moments before interrupting. The two would meet again several times in the next two years, the last being when Reagan learned that Cardinal Cooke was dying of leukemia. The cardinal had tried to keep his ailment a secret, but his failing appearance betrayed him.

When word of the cardinal's condition reached the White House, Reagan called me and said he wanted to see him in New York. After clearing the schedule, I joined the Reagans at the chancellery. The visit was unannounced, not by my design, but at the direction of the attending physician who wanted to avoid exciting the aging cardinal. We attended a Mass and then went to the cardinal's quarters, where he was to give the benediction. After everybody settled into their chairs, Cardinal Cooke beamed at Reagan. "Mr. President, what are you doing here?" he said. "You're far too busy to be spending time on an old man like me."

Reagan replied, "Well, you weren't too busy to see an old man when you came to see me."

Cooke smiled. "I pray for you every day. When I get to heaven . . ." he stopped to collect himself, then offered a correction: "Now that's a bit presumptuous of me, isn't it?" They shared a deep laugh, and the cardinal didn't finish his sentence.

He died within days. When Reagan was told

of his friend's death, the president's words from their earlier meeting echoed in my mind. "Whatever time I may have left is left for Him." I would never forget his promise, and I would see him deliver upon it time and time again.

Reagan's strength gradually returned. He was buoyed by the good news that those wounded during the assassination attempt survived. With the exception of Jim Brady, all enjoyed full recoveries. But even Brady had improved far beyond what the doctors predicted.

One of Reagan's first public appearances after the shooting was at Notre Dame University, where he received an honorary degree. The president drew strength and vigor from the sustained standing ovation. As he stood in front of the crowd, donning cap and gown, I could tell he was happy to be there . . . to be anywhere.

Back in Washington, Reagan addressed Congress, walking to the podium amid a tumultuous welcome. The public was responding with overwhelming sympathy and admiration to his brush with death. But I believe it was his absolute refusal to look back and the gusto and humor with which he went about his remarkable comeback that gave Reagan what amounted to carte blanche in his first term. This is probably when the first healthy coating of "Teflon" was applied to the Reagan presidency. The term provided a convenient, pseudo-scientific cover for critics who couldn't understand why the public loved the man so much. The public saw what the media and the hand wringers couldn't or wouldn't see: a seventy-year-old man who thought that nearly dying was no big deal. He had more important things to do, like live.

Reagan would seldom talk about the attempt on his life, but he was often prodded by high-level visitors to say a few words about that day in March. One of the more memorable exchanges was with a concerned Prince Charles, who wanted to get all the details from a reticent Reagan.

❧

"So, I understand that it was a too-too?" asked Charles, sounding exactly like a prince should sound.

"Come again?" said Reagan, leaning toward his royal visitor.

"It was a too-too, right?" Charles repeated, more animated.

I interrupted, having at last deciphered the prince's "British" that was so alien to Reagan and me. "Mr. President, the prince is wondering if you were shot with a twenty-two-caliber revolver."

The prince nodded enthusiastically as Reagan buckled over in silent laughter. Yes, he confirmed, the prince had identified the right handgun.

I was delighted with—and in awe of—the president's comeback, but I had my doubts about the pace of his recovery, particularly one night that spring, a few months after he had returned to the White House.

Reagan had gone to the Oval Office for a foreign policy briefing. After it was over, I walked in and was startled to see him alone, slouched in his chair. His reading glasses were nearly falling over the tip of his nose. For the first time in my life, Ronald Reagan looked old to me. He was gray, gaunt, and unresponsive. Without acknowledging me, he automatically began to gather his papers when I entered. One fell to the ground, but he left it there. As he began to get out of his chair, I moved toward him to offer a hand, but he straightened his legs with a quick burst, the chair slid backward, and he was on his feet. He told me that he was going back upstairs through the Rose Garden. A cold burst of fear hit me as Reagan hobbled toward the French doors of the Oval Office. He paused there for a moment, looked down, then stepped out into the chilled air. As soon as his right foot hit the pavement outside, he immediately straightened his back and expanded the chest. All of a sudden, the gray pall hanging over him seemed to disappear, and his cheeks colored. All along the passage from the Oval Office through the colonnades to the private elevator that would take him to the second-floor residence, the president would be watched and accompanied by Secret Service agents, but I wasn't thinking about any of that then. All I was seeing was his old verve as he set off on his short walk, and, boy, did it ever look great.

At that moment, Ronald Reagan seemed to pass a threshold, in his own recovery and with me, too. I silently cursed myself as I watched the spring return to his step. I had been working beside him for fifteen years by then. How could I ever have again doubted the man's ability to step up when the pressure was on? He had done it a million times before. I couldn't help feeling like Saint Peter after hearing the cock crow for the second time. Never again would I doubt Ronald Reagan.

In his quarters the president began to exercise again. For years in California, he had done regular workouts under the direction of a physical therapist. Now, his weight and chest expansion were understandably down. He had a universal gym installed at the White House and brought in a trainer from Los Angeles to advise him on weights and routines, and he was religious about his workouts. In a month, he had to have his suits remade. He went up a full size in the chest, and the triceps in his arms seemed to double.

Meanwhile, the question of protecting the president in a violent society was raised anew. I knew there was simply no way to guarantee his or any president's safety against someone crazed enough to attempt a kamikaze mission. This wasn't a good enough answer for Nancy, so we tried to do a few things from my end to ease the Secret Service's burden. We did what we could to limit his exposure to crowds. I never forgot that he had been hit while turning in the direction of a reporter's question. Trying to stop a newsman from shouting a question at an American president was like trying to nail Jell-O to a wall, but from then on, I liked it much better when the president did not stop to entertain questions. Also, we would not again allow the president to arrive or depart from events in a "noncovered" area. He would enter and exit the limo only in a garage or under a tent. There would be no more open exposure, no more clear shots for Hinckley copycats. The Secret Service made sure that an agent roamed among the journalists when they stood as a pack behind rope lines. The Secret Service is the best in the world, but even the best can improve and that's what happened. In short, everybody was much more organized.

I took a lot of heat over the years from my friends in the media about Reagan striding briskly from his helicopter to a car, signaling with his hands to his ear that he couldn't hear over the noise of the chopper blades or stop to answer questions. It was worth it though. Sorry, Sam.

Upon reflection, maybe Edmund Morris was right in at least one regard about Reagan never being the same after being hit with Hinckley's bullet. It took me a while to realize it, but he did change.

After the assassination attempt, Reagan spent considerable time thinking about ways to deal with the Soviet leadership. Nobody has ever doubted Reagan's almost single-minded focus on defeating communism—see his famous "Evil Empire" speech—but he would later make tactical changes in his dealings with the Soviets that would give many conservatives heartburn.

During his convalescence, he wrote a heartfelt letter to General Secretary Brezhnev that he pulled from his bathrobe pocket during a national security meeting at the White House. The letter was six pages long, handwritten on a yellow legal pad. My guess at the time was that Reagan had been touched by the fact he had come very close to death, and what he wrote to the Soviet leader came from his soul. I can't judge how much energy it took the president to write a six-page letter by hand, but I was impressed that a man in his condition would undertake such a project.

He began the letter by reminding Brezhnev that they had met once, at Nixon's estate in California. He went on to say how much the world needed peace and that the two of them needed to think through the historic responsibilities they shared. There were no proposals, just a direct and personal message to jump-start a dialogue. Reagan passed the letter around the room asking for comment. "I don't know whether you fellows think it's a good idea," he said. "But why don't you read it and get back to me."

When the letter was returned a few days later during a meeting of the same group of senior aides, it was a completely different document—neutered of its original intent and prose, replaced with canned State Department boilerplate. We were each given a copy of the letter to review. Reagan looked up from his copy, and said weakly, "Well, I guess you guys are the experts."

I could not restrain myself. "Mr. President," I said, "nobody elected anybody in the State Department or the National Security Council. Those guys have been screwing up for twenty-five years. If you think that's a letter that ought to be sent to Brezhnev, don't let anyone change it."

He turned to Secretary of State Alexander

Haig and said forcefully, "I agree with that. Send it the way I wrote it."

Everybody filed out after the meeting, leaving me alone with Reagan. He thanked me for saying my piece. "You know," he said, "I came to a conclusion about something when I was in the hospital. There must be a reason why I was spared. From now on, I'm going to follow my own instincts."

Already a stubborn man in so many ways, Reagan grasped even tighter to his core beliefs after that. Nowhere was this reinvigorated stubbornness more evident than in his increasingly unpopular stand on the arms buildup.

During the campaign against Jimmy Carter, Reagan had rightly called for a dramatic defense boost, arguing convincingly that years of neglect had sapped the military's morale and reduced our strategic advantage over an increasingly bellicose Soviet Union.

The economy, though, was still smarting from the Carter recession. When it failed to bring in the revenues Reagan needed for his defense increases, he was forced to look at reductions in domestic discretionary spending. His economic and political team, including me, argued that we couldn't cut any more social spending while increasing defense spending with record peacetime boosts. Budget chief David Stockman and others who knew numbers better than anybody argued passionately that massive deficits would result if Reagan insisted in moving forward on defense. Looking back, we made the wrong arguments to Reagan. Forgetting the lessons Nancy Reagan had so deftly taught me back in Sacramento, I cautioned the president that he would take a political beating if he continued to push for more military spending. In hindsight, it wasn't just the wrong argument to make to Reagan; it was the wrong argument to make absolutely.

Defense Secretary Caspar Weinberger was a pillar at Reagan's side during this battle, but he had little company within the cabinet or elsewhere in Washington. The Republican leadership on the Hill was quivering; the elite media were pummeling him without mercy; the

Democrats were doing what Democrats are supposed to do. Worst of all, the American people were dead set against the Reagan buildup. The country was limping through a brutal recession, and Reagan's approval ratings felt the country's wrath.

Instead of succumbing to self-pity or changing policy, Reagan pulled me aside one day. "Mike," he said, "these numbers show you're not doing your job. This is your fault; you gotta get me out of Washington more so I can talk to people about how important this policy is." I did, and he would systematically add his rationale for more military spending to nearly every speech, and eventually his message would get through to the American people.

Reagan would say time and time again that if given a choice between big deficits and winning the Cold War, he'd take the latter every time. I always marveled at his confidence when he said, "I'll take full responsibility for this gamble." The Reagan presidency didn't get to enjoy the full economic benefits of the collapse of the Soviet Union—that would fall mostly to the Clinton years—but history has a way of eventually sorting through the debris and finding who counts. And history, I'm convinced, will be kind to Ronald Reagan on this matter.

Ronald Reagan was a man who knew where he stood. The attempt on his life did nothing to change that, but it did make him more confident in the judgments he would make as commander in chief, peacemaker, and head of state. And for that, maybe, in a weirdly twisted way, the nation has John Hinckley's twisted mind to thank.

CHAPTER FIVE

A Guy Named Ron

What's the secret of Ronald Reagan's immense success? The guy knows who he is. In the public relations business, I constantly tell clients—corporations or candidates, it doesn't matter—if you don't know who you are, there's not a lot I can do for you. My first realization of how comfortable Reagan was in his own skin came during the late 1960s. We were staying at the Waldorf-Astoria Hotel in New York City, where the governor had given a speech. The next morning, I received a call from him. "Let's go for a walk," he said.

I met him in the lobby fifteen minutes later and we strolled out onto the sidewalk. I could tell Reagan was feeling chipper by the look on his face and the spring in his step. Within about ten minutes or so, a middle-aged fellow approached us with a beaming smile. "Hey, I know you from television, and you're the best. You're Ray Milland!"

Without embarrassment, Reagan looked down toward the sidewalk and grinned as the man shoved a pen and paper into his chest, seeking an autograph. Reagan obligingly signed it "Ray Milland." The man walked away happy as a clam.

"Why did you do that?" I said. "You're the governor of California. Why didn't you tell him?"

"Mike," he said softly, "I know who I am. Don't ever worry about me."

I would have plenty of cause to worry about Reagan in the decades ahead—cause even to fear for him—but about his sense of self, his understanding of who and what he was, I would never again fret. Lots of people in politics are someone else's man; that's the nature of the beast. But Ronald Reagan belonged to no one else. He believed in bedrock values: honesty, humility, and integrity maybe first among them. And he never aspired to be anyone or anything other than himself. In fact, when I think back over the twenty years that I worked for Reagan, and the countless people who came in and out of his life, the only one who didn't change was Reagan.

That doesn't mean he wasn't complicated. Reagan had any number of traits that generally flew below the radar of the public and the media. For one thing, he had a hungry intellect that even Clark Clifford might have appreciated if he had taken the time to get to know the president. Despite his reputation for a laissez-faire approach to his duties, Reagan was also both extremely competitive and extremely disciplined. As I've written earlier, he left the details to others, but no one worked harder at mastering the big picture, and America is the better today because he did.

Nor does the consistency of Reagan's character mean that he was easy to get to know. So frustrated was Edmund Morris in trying to unlock the secrets to Ronald Reagan that he resorted to fictional creations in his biography *Dutch,* the childhood name by which the president was known back in Illinois. I can't imagine what it's like to try to capture a life as sweeping as Reagan's in a single book; even those who worked for years with him often found a final wall they couldn't get through. Reagan was an acquaintance to many but intimate with only a few. But I think the problems his biographer and even a number of his aides and associates have sometimes had in understanding Reagan and feeling a sense of closeness with him are rooted in what Stu Spencer first noted back in the 1966 gubernatorial campaign: Ronald Reagan is one of the shyest men I've ever known.

In a let-it-all-hang-out age, America found itself being led, ironically, by a man who really couldn't stand to talk about himself.

A friend and colleague in the administration, Ken Adelman, once hosted a dinner party for President Reagan at his home. Naturally, Adelman's invitation list was a who's who of

Washingtonians—senators, cabinet secretaries, congressmen, media elite. He also decided to invite his father, a retired laborer. By chance, the senior Adelman was the first person Reagan met that evening. Through the course of the party, Reagan never left his side. God knows what they talked about, but he was perfectly happy talking to this regular guy in a room full of national leaders.

We used to try to guess the genesis of Reagan's reticence. The fact that his parents, Jack and Nelle, relocated many times chasing new job opportunities probably had something to do with it, but I also think his social development was affected by the fact that he was almost blind until his parents could afford to buy him spectacles. In fact, until a visiting nurse came to his grammar school in central Illinois and tested his eyes, Reagan had been considered a slow student. The nurse was the one who realized his "learning disability" was severe nearsightedness. Reagan told me that until he got his glasses, he lived in a closed world. "I never knew there were leaves on the trees," he said. "I never knew what a butterfly was. I could never see the blackboard and was never picked for a team until I got my glasses."

His vision problems early in life certainly influenced Reagan, but I can't help but think they're only part of the story. After all, I've always felt that however well he can see, Reagan would be just as happy in an empty room as a crowded one. Where does that come from? Well, I have a theory there, too, this one deeply personal.

It is commonly known that Ronald Reagan's father, Jack, was a chronic drinker, an alcoholic. Children of alcoholics are usually kindred spirits, cut from the same cloth and able to identify with each other. More often than not, sadly, the disease is passed through the generations, afflicting not only your children but frequently your children's children.

I know a little more than the average guy about the disease of alcoholism. It may be the most profound thing about me. I grew up in a house with alcoholic parents, and alcohol ultimately ruled my life until 1986 when I checked myself into Father Joseph Martin's Ashley Center for Alcohol and Drug Rehabilitation in Havre de Grace, Maryland. Father Martin, a recovering alcoholic for more than forty years, has influenced and helped thousands of alcoholics and their families. He saved my life and I'll be forever grateful.

In his childhood and early adolescence, Reagan's father lived to drink. Frequent "business" trips to the big cities were nothing more than drunken binges. The story has been told so many times that it sounds like a scene from a Frank Capra movie that Ronald Reagan might have acted in, but it's true all the same: one Christmas, "Dutch" found his father passed out in a snowbank, quite literally freezing to death, and carried him to bed, almost certainly saving his life. I'm sure there were other similar tales—if you can ever get us going, we children of alcoholics can tell stories like that till the sun comes up—but in the countless hours I spent with Ronald Reagan, nearly every day for twenty years, he never really talked about his father's losing battle with the bottle. He never used it as common ground to start a conversation. He never used it to comfort a person fighting the disease. Fortunately, Ronald Reagan never had the disease our fathers had. Reagan didn't inherit the gene; I did.

Not that I didn't try to get him to open up about his father's drinking or, more important as I saw it, Reagan's lack thereof. When we did talk about alcoholism on two or three occasions, Reagan would tell me the same thing each time. He would detail the amazing grace demonstrated by his mother during those tough times. After lugging Jack into bed after one of his binges, Nelle would sit young Reagan down at the breakfast table the next morning and explain that his dad was suffering from a disease. "It was not his fault. Nelle was right all along," Reagan would tell me.

Nelle and Jack. I never heard Reagan call them Mom or Dad, Mama or Pappa, or anything else. It was always Nelle and Jack. Puzzled, I would try to dig for an explanation, but as with so much deeply personal to Reagan, I'd come up with a blank there, too. "That's what they wanted me to call them," he'd say. End of story.

Nelle Reagan's insight and tolerance are remarkable considering the period in which she lived and the lack of medical insight into the disease at the time. Her foresight and grasp must have been critical in helping Ronald Reagan cope with what he saw. But even with such an understanding mother, one of the lessons a child learns early on in an alcoholic family is to keep his own counsel, to stay silent about some of the things and people closest and dearest to him. Ronald Reagan, I suspect, learned that lesson in spades and has never forgotten it.

The other story Reagan would tell me when on the subject of drinking was about how he played a drunk during his acting days. Reagan, a dedicated student of the stage, sought out a real-life drunk as a consultant, just to make sure he had it down. At one point, he engaged the man after a practice session, asking why he felt the need to drink so much and if it were possible for him to kick the habit. "If I could just avoid the first drink . . ." came the response. Reagan thought this was Gospel. "If alcoholics could just learn this lesson, they'd be okay," he would later tell me, totally convinced. It was pure Reagan—this impulse to reduce the complexities of life down to easy axioms—but as simple as this observation seems, it took me years to realize that he was absolutely right.

I never heard him talk about anybody else's drinking. Never. I never heard him talk down about somebody that wasn't blessed with his discipline. The closest he ever came to commenting on my alcoholism was back in our Sacramento days when the two of us were entertaining guests one night. I was playing the piano and drinking; Reagan shared the bench with me, singing away. We stayed up well past midnight. A week or two later, I received a photo of Reagan and me at the piano with the inscription: "Dear Mike, One more chorus of 'Springtime in the Rockies,' and this time use the black keys too."

He did tell me once that there was a time when he enjoyed martinis. "I started to enjoy them too much," he said, "so I quit drinking 'em." Apart from his glass of red wine with dinner and an occasional—and I mean *occasional*—screwdriver, Reagan's discipline held as long as I knew him. I never saw him crave a martini, a gin and tonic, or anything else. Needless to say, I never saw him drunk, tipsy, or out of control. He didn't need the stuff.

He didn't need or have any tolerance for drugs, either. Governor Reagan had precious few fans on America's college campuses as the Vietnam War escalated and the protest and hippie movements grew. His support for the war and his policy of zero tolerance in California for campus scofflaws had caused him to be burned in effigy countless times and had brought in numerous death threats as well. Still, Reagan wasn't one to be cowed, and he always agreed to visit campuses or meet with protestors.

It was during one such appearance before faculty members and students at Vanderbilt University that an unyielding heckler interrupted Reagan a third time.

"Hey Ronnie, what do you think of pot?" the provocateur shouted. His question reverberated in the auditorium to great, prolonged laughter.

Reagan waited until there was silence and said, "Well, I think you know my answer, but you're not going to like it."

"Your generation needed your martinis, right, Ronnie?" the invisible protestor shouted again.

Without batting an eye, Reagan said sincerely, loudly, "You're right, we did need our martinis, but I can still hope that a generation would come along that wouldn't *need* anything." Maybe if the heckler had lived through what "Dutch" Reagan lived through growing up, he would have understood better the governor's aversion to anything that robbed you of self-esteem and self-control. Who knows, he might even have agreed.

Reagan has a fierce competitive streak, which I believe had its genesis in Hollywood, where beating the other guy was not just a way of life but was required for survival. "One-Take Ron," as he was called by many of his appreciative directors, would never be unprepared for a shoot, staying up all night when needed to learn his lines to perfection. It was just one way he got an edge on the competition.

A legendary story, set during his heyday in Hollywood, has Reagan filming a beach scene with Errol Flynn, a contemporary and friend. Reagan knew that the taller Flynn was Hollywood's hot ticket and had enormous star power. A shot of Flynn looking down on him was the last thing Reagan needed. He would not be comfortable being perceived as Flynn's junior.

For some reason, to the director's chagrin, the scene required four takes to complete. One-Take Ron kept flubbing his simple lines. Later, when reviewing the dailies, something unusual was noticed. The first take showed a shorter Reagan looking up to Flynn as they engaged in their discourse, but by the fourth and final take, Reagan seemed to grow to a point until he was standing over the dashing Flynn by about an inch or so.

Did Flynn become fatigued and slouch? No, he stood proud as ever. But starting from the first take, Reagan used his feet to build a sand platform to give him more height. During each take, he'd add to this makeshift pedestal until he was in the enviable position of appearing to be Flynn's equal.

The Flynn tale aside, One-Take Ron truly earned his nickname. He practiced extra hard not only because of the competitive beast that lurked inside, but also because he had been taught the value of self-discipline, mostly by his mother, Nelle. It was evident in nearly every part of his life, from the way he packed his suitcase—not wasting an inch and making a list Santa would be proud of—to how he dieted and exercised. He must have given the same campaign speech ten thousand times, but he practiced it in its entirety before each session. He would not be caught flatfooted.

If he were still alive to comment, I feel fairly

certain Bobby Kennedy would agree. Like a lot of sophisticates, Bobby underestimated what Clark Clifford would later call this "amiable dunce," and he paid a heavy price for his error. The occasion was an intercontinental debate held early in 1968. Taking advantage of nascent satellite technology, Reagan, then the sitting governor of California, and Bobby Kennedy, well into his campaign for the presidency, agreed to debate on the major issues facing both parties. Although Reagan was not a candidate for president, he was nationally recognized as a conservative luminary. And Kennedy, well, was not just a candidate but also a Kennedy.

Reagan was in Sacramento and Bobby in Paris for some reason. Despite the distance, they engaged in a spirited debate, Reagan making point after point on everything from taxation to foreign policy. Naturally I thought Reagan got the better of Kennedy, but I never really gave the session much thought until about twenty years later when I was at a charity dinner in Boston. There I met an attorney who had been on Bobby's staff at the time of the debate. When the lights went out in the studio in Sacramento, Reagan looked at me and smiled. He didn't say anything and I don't think the debate ever came up again. It was a different story in Paris according to the Kennedy aide. Bobby looked at him and said, "Don't ever, ever put me on the same stage with that son of a bitch."

Reagan had the smarts; he just wasn't a showboat. He craved reading, a love that he got from his mother. In all my years with Reagan—on the campaign plane or in the White House residence—I watched him devour countless books, white papers, position papers, memos, and biographies. Reagan once told me on a campaign plane, "Mike, Nelle told me that if you learn to love reading, you will never be alone." I later understood this to be the adage of a shy man, and Reagan lived it.

His briefcase was always full of things to read when he was able to get free time. Yet Reagan didn't consider himself an intellectual, and he didn't care if anybody thought he was or wasn't. The smart guys could fill the op-ed columns with all the bombast and erudition they could muster, but the first thing Reagan would read every morning was the funny pages of whatever the local paper happened to be. He liked to start the day off with a smile and would try to share some laughs with the people he was with. In Sacramento, *Los Angeles Times* reporter Dick Bergholz spent more time around the governor than most of Reagan's aides. He was sometimes a dour fellow, but whenever Reagan was with him in the early hours, the governor would try to tweak him into glancing at the funny pages in hopes of turning Bergholz's frown upside-down. It usually didn't work.

When it came to the media, Reagan didn't have to worry about fudging the truth or skirting the issues; neither was a part of his makeup. Sometimes during interviews, a reporter or aide would caution grimly, "Remember, sir, this conversation is on the record." The warning was a complete waste of time. It made no difference to Reagan—he was going to say the same thing whether the conversation was for just a few ears or for everyone. I would go so far as to say that he was actually incapable of dishonesty in any form. Not just when it came to the big things—there was never any question of that—but even when tested by the everyday opportunities we have to fudge things just a little.

In the late 1960s, President Nixon asked Reagan to represent him in a series of meetings in the Far East, and several of us who worked for the governor got to go along for the ride. Needless to say, great shopping opportunities presented themselves at every stop. As we were heading back to the States, the purser on the military aircraft handed out U.S. customs declaration forms but advised us, "Not to worry folks, we'll probably receive 'courtesy of the port.'" Translation: we might not have to go through customs, but there was also a good chance we would get a free ride and not have to answer to Uncle Sam for the many items we picked up on our "shopping" excursion. Our military liaison urged us to fill out the customs forms just in case.

All this caused considerable angst and gnashing of teeth. I remember consulting with my fellow travelers whether we should mention the many Asian suits and ties and shoes that we bought. After all, if we were going to get a free pass, why bother? On the other hand, there was an outside chance the customs guys would want to check us out.

In the end, we were waived through. After the Reagans left the plane, I stayed behind to collect some of my papers from the meetings I attended. I also took note of the dozen or so customs forms lying about. Most of them were fiction that would make J. K. Rowling proud. Then I came across Ronald Reagan's completed customs form. He had listed every single trinket, garment, and souvenir he purchased over the course of the entire trip. Some of them were literally worth pennies. It was a remarkable piece of bookkeeping, I thought. But even more impressive to me was his total and complete honesty.

It simply didn't occur to him that he could probably slip under the radar duty free. Looking at his form, I calculated that he would actually have had to pay a small duty for the goods he bought. Sure, the money wouldn't have worked a hardship on the Reagans. Getting caught wouldn't have been that big a deal, either. Let's face it, fudging customs statements is one of America's most popular sports. But there's an old saying that grows more true to me with every passing year: character is what you do when nobody's looking.

We used to laugh about a story that happened when Reagan was governor of California. My daughter, Amanda, grew up as a toddler knowing that I worked for the governor and that he was in a position of high authority. When Richard Nixon gave his speech in the East Room of the White House in 1974, finally resigning the presidency, Amanda watched her mom crying in front of the television before finally asking why. All Carolyn could say was "Because the president lied, honey." That seemed to suffice. After all, how do you explain to a three-year-old that a man has violated his oath of office and for the first time in the history of the United States the man elected to its highest position has been forced to resign in shame?

The next time Carolyn brought Amanda to the governor's office, a few days later, she ran directly to Reagan. He scooped her up, and to everybody's surprise she began to cry and shake violently. Reagan took her into his office and closed the door behind him. Carolyn and I just stood there, staring at each other. Five minutes later, she walked out with a handful of jelly beans, her eyes completely dried.

After Amanda left with her mother, I asked Reagan, "What was that all about?"

"Beats me," he said with a shrug of the shoulders. "She just kept asking me, 'You'll never lie, will you?' I told her I wouldn't, and that she shouldn't worry."

* * *

Reagan's roots were Irish, but he wasn't really an emotional man. He had an unabashed love of country that would often express itself in emotional ways, and of course, his love of Nancy was—and is—unconditional, too. I know for a fact that he lost it when the White House physician broke the news that she had breast cancer. Otherwise, I almost never saw him succumb to tears in the sort of situations that would reduce the rest of us to jelly, but that doesn't mean he didn't feel things deeply.

A scene that touched me the most during my five years at the White House was one few saw or ever knew even happened: a private meeting in the Oval Office with the Polish ambassador to the United States, who had just defected from the communist regime, and his wife. The ambassador sat in a leather chair next to President Reagan. George Bush sat next to the ambassador's wife on the couch facing them. I stood in the background.

The ambassador was referring to notes as he spoke. Having just abandoned family and friends to establish a new life in the free West, his wife wept silently into her hands as Bush tried to comfort her.

"It is unbelievable to me that I am sitting in the office of the president of the United States. I wish it were under better circumstances," the ambassador said. His voice quivered as he read what he thought were his most important words. "You must never end Radio Free Europe. You have no idea what it meant to us to hear the chimes of Big Ben during World War II. Please, sir, do not ever underestimate how many millions of people still listen to that channel behind the Iron Curtain."

Reagan was visibly moved but remained silent. When something got to the Gipper, he would have a pained look on his face, almost like a grimace. The ambassador then asked the president to light a candle for the people of his homeland and put it in a White House window.

I followed Reagan and his Polish guests as he escorted them out the walkway to the circular driveway on the South Lawn. In a driving rain, he took them to their car, past the Secret

Service post. The entire time, he held the umbrella over the woman's frail head, which rested on the president's shoulder, as she began to sob uncontrollably.

Reagan watched in solitude as the car took them away. When he walked back, he had the look of a drained man. His eyes were swollen, but he said nothing. Later that evening, in the second-floor residence, Reagan lit a small white candle and put it in the window of the dining room.

When it came to stuff like this—America and freedom and liberty—Reagan was a softy. And his belief that America was the greatest country on earth never once wavered even when those around him refused to share in his idealism. After a campaign stop as governor of California, Reagan and I were alone at twenty-five thousand feet in a luxurious private jet. We had just finished a gourmet meal, drunk a glass of fine California red, and were about to attack a sinful dessert when Reagan looked out the window wistfully.

He said, "Mike, you know the great thing

about America is that anyone of those guys down there can be up here if he really wants to."

I was almost taken aback by his naïveté, but noticing my grimace, he persisted. "What I mean is, don't you want to live in a country where it's at least possible?"

Sadly, he had to bring himself down to my level to make his point. I almost felt guilty for ruining the moment, but I later thought, isn't it better to have a man like this leading us than somebody who thinks the opposite?

Reagan loves to tell stories of average Americans. One of his favorites was about a group of Ohio tourists who were visiting Mount Vesuvius in southwest Italy. After hiking to the top of the volcano, one of the elderly tourists, a portly man in his late fifties, leaned a little too close to the volcano's mouth to get a closer look. As he stared at the smoke emitting from the ancient opening, his Italian guide cautioned him sardonically in broken English: "Better stay away, sir, that thing could blow any minute."

The American looked at his guide and shrugged. "Hell," he explained, "we've got a volunteer fire department back home that could put that thing out in twenty minutes."

Moments like this came easily to Reagan. He was red, white, and blue through and through, and he never minded showing it, whether patriotism was fashionable or not. When it came to more personal matters of the heart, though, Reagan was often accused of checking his sentimentality at the door. Worse, these charges often came from those he loved most, his children. During the early summer of 2000, I was watching the second airing of a very balanced two-part series on the life and presidency of Ronald Reagan that had been done for the excellent PBS series *The American Experience*. The producers interviewed some of the Reagan children, and the interviews were telling.

One moment that especially caught my attention came during a conversation with Reagan's son, Ron. Ron was bemoaning the fact that he never really felt that close to his old man. Sure, his father loved him, Ron allowed, but he wasn't ever there for him in an emotional way. By way of illustration, he went on to tell a story about his early days as a dancer.

Ron accurately recounted a time when I pulled him aside on the campaign trail to tell him how proud his dad was of his new career on the stage. But he got it wrong when he told the PBS viewers that the president actually directed Mike Deaver to make this overture. Sorry to ruin a good yarn, Ron, but it ain't true.

I never once was asked by either Nancy or the president to do anything like act as a surrogate for an emotionally distant father. I had simply overheard Reagan bragging about Ron prior to a meeting with some visitors. They had seen a picture of the family, and Reagan volunteered that young Ron was moving in a new direction. He beamed with pride at his son's independence and ability to handle the glare of the spotlight.

I had passed this tidbit on to Ron. I probably should have kept my mouth shut, but I thought it might help demonstrate how much

Reagan really cared for his son. Back when Carolyn and I first got to know the Reagans, in Sacramento in the mid-1960s, when Ron was just heading into his teenage years, their family life seemed about as normal as it could be for people who lived in a governor's mansion. There were birthday parties for Ron, touch football games on the lawn, the everyday stuff of family relationships. As the years went on and Ronald Reagan became a national and later international figure, father and son spent less and less time together, to be sure. Ron was growing up and moving on, too, and Ronald Reagan was a father that all his children eventually would have to share with the American people, even, in a sense, with history.

It's very difficult to understand, but I rarely heard Reagan say thank you to me. That's not really the strange part though. The odd part is that I didn't care. None of his staff ever talked about it or felt a need to seek his praise. It never once bothered me or any of my colleagues.

Whatever we did for Reagan we did out of

loyalty to the man. I did my work because I believed in him and what he stood for. He wasn't the type to pat you on the back, but you knew when you did something right. You always knew instinctively where you stood with Reagan, and the communication channel ran both ways. When Reagan would make mistakes, I'd tell him without varnish.

During a campaign stop in Chicago in 1980, Reagan delivered a humdinger of a speech. It was just before the evening news cycle and was sure to be the lead story. I pulled Reagan aside after the speech with some other aides to let him know how strong the speech was and to suggest as subtly as we could that nothing else be done that evening to upset the apple cart.

Reagan nodded in agreement. "So you're saying I should keep my mouth shut?" he asked. We nodded.

Reagan agreed that we would walk silently to the waiting car, but for some reason only God knows, as soon as we got there Reagan stepped up on the running board and began going at it with the press vultures, saying just enough to contradict the wonderful story we had just put a bow on.

He got into the car and his silence told me that he knew he dropped the ball. "I don't believe you," I said, probably too loudly. He said nothing as we drove alone toward the next stop.

"I just don't believe you," I said again, louder.

This time, he turned toward me. "If you're so damn smart, why aren't you running for president?" I took it as a rebuke, but although his tone was on the sharp side, there was that characteristic twinkle to his eye.

We didn't say another word to each other for the rest of the ride or during the two-hour flight to the next stop. That night as I was preparing for bed in another forgettable hotel, there was a knock at the door. It was the governor's personal aide. "Mike, sorry, but the governor wants to see you," he said.

I walked into Reagan's suite. He was wearing a robe and was looking down at a small rectangular box that he was fiddling with nervously. When he opened the box and pushed it toward me, I saw a gold Cross pen that he must have received earlier in the day as a gift.

"Mike, I thought you'd like to have this."

"Why would I want that?" I said. "You must have given me a dozen of those things over the years."

"Well, maybe your son, Blair, would like to have it?"

I wasn't going to let him off easy on this one. "What do you think I did with the other twelve?"

He frowned and shoved the pen in my hand and turned around. The word "sorry" was never uttered by either of us, but it wasn't important. I knew he had just apologized to me and the dance was done. It was time to move on to the important stuff. The next day all was well. We never spoke about the Chicago blowup again.

The words he did say and the way he said them were often memorable. Reagan clearly knew that his soft, inviting tenor was his bread and butter, in large part responsible for Reagan's success in politics, business, and film. Most great orators of our time wave their arms or gesture majestically with their hands. Clenched fists fly in the air, and fingers point like piercing daggers—but never Reagan. He didn't need his body for emphasis. The pure resonance of that voice was more than enough.

By way of illustration, you need look—and listen—only to Reagan's performance at the Berlin Wall when he inspired the Free World with his call to arms: "Mr. Gorbachev, tear down this wall." Given the same opportunity, JFK had used his fist as a way to drive home his message that he stood with Berlin, but Reagan grew up in the FDR school where radio was king, not television, and he knew that words alone were enough when you delivered them right. I had nothing to do with creating the moniker the Great Communicator, but I do know that Reagan ranks with FDR and JFK as the only three presidents of the twentieth century who could move the country with what they had to say. After all, Reagan was a performer. Aside from his voice, the confidence and timing are both there, honed by thousands of speeches and scripts.

When considering a radio career over television after he left the governor's office, Reagan told me that radio was a better fit. "People will tire of seeing me," he said, "but I don't think they will tire of hearing me." In fact, they didn't tire of either. Whatever problems of intimacy he suffered, Reagan connected countless times with people he had never met before and would probably never talk to again. In these relationships, Reagan excelled.

Throughout his political life, Reagan insisted on getting a sampling of the mail. He'd personally answer every one of these that made it to his desk, and not always just with words. When he was governor of California, I walked into his office early one afternoon to go over the next day's schedule only to find him clearing his desk, a clear signal that he was closing up shop for the day.

"Leaving so soon?" I asked.

"Yeah, got a few errands to run," was the reply from the governor.

From time to time, this would happen, and naturally I was always curious. This time I pulled his driver, Dale, aside and told him I'd want a full dump on what happened in the morning. After Reagan left, I looked through his "to read" file that sat in solitude on the otherwise empty desktop. The first piece of paper was a wrinkled letter from a man stationed in Vietnam. The soldier, a resident of nearby Orangevale, had written a letter to the governor telling him about life in Southeast Asia and how much he missed his wife. He was miserable and wanted to be able to tell his wife how much he loved her and how much he longed for her. Today was their wedding anniversary, and he would be without his soul mate. Although he wrote that he had already sent the obligatory Hallmark card, the soldier asked if Reagan could put in a phone call to make sure she's okay. He wanted the governor to pass on his love if she didn't receive the card.

The next day, Dale told me that Reagan had done more than that. He'd picked up a dozen red roses and delivered them to the soldier's wife in Orangevale personally. Dale recounted that the governor approached the woman in a grandfatherly way, humbly offering the flowers on behalf of a loving husband stationed in a jungle hell on the other side of the world. Afterward, he spent more than an hour with the woman sipping coffee and talking about their kids.

The next day, when I asked Reagan how the errands went, he said with a smile, "They went fine, fine."

Another time, while I was sitting in the governor's office with Reagan going over some business, the door cracked. It was Dale. The governor smiled and waved him in with a hurried sweep of his hand. Dale was carrying a plastic garment bag that he opened to reveal a suit—one of Reagan's flawless blue gems, probably costing in excess of two thousand dollars. Reagan inspected it, then felt the sleeve and nodded approvingly. Dale solemnly walked out.

When I asked what was going on, the governor passed me another wrinkled letter. This one was a handwritten note from an eighty-year-old man named Hennings. It read something like this: "I'm getting married soon and I want to look like you. I've seen you on the TV a few times, and we look about the same size. I'd sure appreciate any help you could give me in getting a new suit. I want to look my best for my big day."

I jumped out of my seat in hopes of catching Dale. If we really had to do this, I called back to the governor, I'd have somebody pick the old guy up a Haggar suit from J.C. Penney. Reagan would have none of it.

"I've already talked to this Hennings, and we actually are the same size," Reagan said, stopping me. "And he's expecting this now, so let Dale take care of it."

I don't know if the governor was invited to the wedding or not, or if "Hennings" ever got back to him with a thank-you note. It didn't matter to Reagan. He lived up to his part of the bargain, but it wasn't like he had clothes to burn.

In 1973, Reagan was in Los Angeles for a gubernatorial function. I stayed behind in Sacramento to tend the store. It was a quiet Saturday morning, and I was just about to relax with the paper and a cup of coffee when the phone rang. Carolyn took the call from Governor Reagan, then summoned me with her hand.

After we'd exchanged pleasantries, the governor gave me the reason for his early morning call. "Mike there's a plane coming from Sacramento to L.A., and I need you to get something on that plane for me."

Was it a speech or policy paper? No. He needed his red polo shirt. The tone in his voice let me know it was important to him. He went on to provide very detailed instructions on where the precious shirt was hiding. "You go up the stairs, take the first right into the master bedroom," he said. When I didn't respond, he asked if I was writing this all down. I assured him that I could handle the task.

He continued, "When you go in my walk-in closet, it'll be on your right-hand side, first shirt from the back."

I had never been in Reagan's bedroom, so I

assumed that there'd be a lot of clothes to wade through. After his maid let me in, I began my ridiculous task, following the directions to the letter. When I marched into his walk-in closet, the red polo shirt was the only thing in the entire closet. Admittedly, he kept the bulk of his clothes back in Los Angeles, but for a guy who looked most of the time like a million bucks, Ronald Reagan was no clothes horse.

After he was elected president, Nancy made a point to upgrade his collection. She discovered a fine English tailor outside of London. Together we selected a blue and gray plaid swatch and sent the order to the tailor, who kept the president's measurements on file. On the tiny swatch, the pattern looked terrific, but as I later found out, the part doesn't always suggest the whole when it's expanded into suits, couches, or wallpaper.

When Nancy eagerly opened the box from London, she was repelled by what she saw: a very large, blue glen plaid that might have been turned into a handsome sport coat or pair of trousers but in this case had been made into a suit fit for a racetrack tout. Nancy was about to deep six the gaudy concoction when the president walked in and proclaimed simply, "I love it."

The suit soon became his favorite, and amazingly he found it appropriate for all occasions. He was wearing it, again, once when he boarded Marine One, the president's helicopter, where Nancy—who had not seen him since he had slipped out of bed that morning—was waiting for him.

We were airborne when Nancy finally reacted. "Ronnie, please, I wish you would give that suit away. If you don't, I may burn it."

He was stunned. "What's wrong with it?"

She looked to me for backup. "Mike, tell him about the suit."

I tried not to smile. "Around the office, whenever you wear that thing, everybody says, 'If he had to be shot, why couldn't he have been wearing that suit?'"

To be sure, Nancy and I were extrasensitive of protecting the Reagan image. He could have been wearing a burlap sack and been just as confident. I should have kept this in mind, but visuals were my game. My wife, Carolyn, had the best answer when it came to putting Reagan the fashion plate into perspective.

It was a Sunday afternoon in Sacramento sometime in the late 1960s. Governor and Mrs. Reagan had invited a few aides and their spouses over for a summer barbecue. The first thing I noticed was the governor standing proudly in a small circle, telling jokes, dressed in a gruesome white wool blazer with gaudy, oversized bronze buttons.

On the way home, I was telling Carolyn how funny I thought the man looked. When she didn't laugh, I began to wonder how she could have missed the outlandish garb. She set me straight, "When are you going to realize that nobody looks at Ronald Reagan's clothes? They only see him."

How right she was, but it took me a while to understand.

It wasn't just clothes and flowers that Reagan would bestow upon people who wrote him; in his heart were many mansions. Another classic tale of Reagan's bonding with a perfect stranger resulted from his reading the troubling account of a welfare mother who was having a hard time getting enough cash together to buy school supplies for her kids.

After reading the article, Reagan, then president and the single most powerful man in the world, wrote the woman a check for two hundred dollars. He included a small note saying he hoped this tiny contribution would help her make it through the tough times she was having. Reagan was later told the check never cleared, so he had his personal secretary find the woman's phone number.

I was in the room as the president of the United States dialed the number. "Hello, it's Ron Reagan," he announced. He spent a good ten minutes on the phone with her, nodding and smiling most of the time. "Okay, that's fine, but you'd better cash this one," he concluded.

It turned out that her neighbor told her to frame the first presidential check because it would be worth considerably more as a keepsake. Reagan understood this and promptly cut her another one, admonishing her that she needed to put this check to good use.

It didn't matter who you were or what your position in life was. Ronald Reagan treated everybody the same. Whether you were a soldier's wife or the Queen of England, you got the same treatment from Ronald Reagan. I was with him in Windsor Castle, and I can testify that Reagan treated Queen Elizabeth exactly the same way he treated White House stewards—with grace, kindness, and respect. He was absolutely the same with everybody.

I have been asked, more times than I can remember, what is the happiest I ever saw Ronald Reagan. Two instances come to mind.

I always liked to joke with the president that the only reason he ran for office was so that he could throw out the first pitch on opening day in a big league park. He never disputed the charge, and I think one of the great disappointments during his first year in office was that the assassination attempt had prevented him from doing it.

In 1984, however, he would not be denied. Opening day at Memorial Stadium, then home of the Baltimore Orioles, was where the president would fulfill the boyhood dreams of many of us. I watched from the dugout as Orioles catcher Rick Dempsey greeted Reagan on the foul line between home plate and first base to loud roars from the festive crowd. It was clear that, after handing the ball to the president, Dempsey was instructing him to take up a position halfway between the pitcher's mound and the plate. Obviously, the poor guy had fielded a lot of lousy pitches hurled by politicos, and he probably wanted to save the Gipper some face. After all, a toss from the mound to home plate is about sixty feet, a tad long for your average septuagenarian.

After Reagan received the ball, he dutifully listened to the unsolicited counsel, all the while backing up toward the mound, nodding in the affirmative every step of the way. Clearly,

Dempsey's advice had fallen upon deaf ears, or more accurately ears that wouldn't hear. Slowly and steadily, Reagan had positioned himself up on the pitcher's mound. The catcher, realizing that he wouldn't be able to talk Reagan into doing anything but the Full Monty, assumed his position behind home plate. The Gipper wound up and dealt a dead strike, an aspirin tablet. I could actually hear the pop of the mitt.

It is probably one of the happiest times I had seen Reagan have. As Dempsey sprinted out to the mound to offer his congratulations and the commemorative ball, he must have been wondering where the old man got his stuff. The answer is the same old answer to so many Reagan questions: He got there through hard work.

Unbeknownst to Dempsey and everybody else in Memorial Stadium, Reagan had spent the previous six or seven weekends practicing at Camp David. During downtime, the president would engage a Secret Service man to help him with his fastball. Ronald Reagan would not be throwing a wild pitch in front of the American people.

The other incident occurred years earlier. A local GOP volunteer had picked the governor and me up at a small southern Illinois airport. On the way to the hotel, I asked our new friend, a local physician, how the fishing was in these parts. He answered that the bass were biting and we had picked a perfect time to be in Illinois if we wanted to catch a few. I looked over my shoulder and noticed Reagan looking out the window absorbedly, not hearing a word.

I wasn't surprised. Reagan had a big speech that night. Although it was another in a long line of political talks he would give, he would spend the remainder of the afternoon practicing his remarks. After we got Reagan settled, the doctor asked me if I wanted to go wet a line in a pond that wasn't too far away. I said yes immediately but told him I needed to check with the boss. When I walked into Reagan's room, he was writing some notes on a three-by-five card. He looked up at me.

"There's great fishing around here, and the guy who picked us up offered to take me out for a couple of hours," I said. "You okay without me?"

He patted his lips with the eraser of his pencil and asked, "Do you think he'd mind if I came along, too?"

I couldn't believe what I heard. The Iron Man of speech preparation would actually throw caution to the wind and try to have some fun. "Let's go," I said.

I thought that we'd be going to some bucolic, Rockwellian place to fish, but to my chagrin, our host took us to his farm and led us to the bank of an irrigation ditch surrounded by cows and the sort of by-products cows are famous for. I was frowning as I looked toward Reagan, but he looked as if he was frolicking among roses, not cow patties. He baited his hook with a smile and took his place among the thirsty cows about fifteen paces from our host and me. Every few minutes I'd look over to see him pulling fish out like one of those guys on the

Saturday morning sportsman shows. He was putting on a clinic.

We ended up spending three hours on the banks of that ugly ditch near a town I can't remember, probably catching twenty or twenty-five fish. I never saw Reagan laugh so much or have so much fun. The smile never left his face. We told endless jokes, lied about bigger fish, and reminisced about our youth. He regaled our doctor-host with stories of his sportscasting days at Des Moines's WHO radio, talked about the hapless Cubs, and probably debated with him about the best way to perform a tracheotomy, too.

About two hours before the speech, I reminded Reagan why we were there in the first place. He looked at me and reluctantly reeled in, and we packed up, eternally grateful for this diversion from the daily grind.

Reagan was nearly always in a great mood, but those spontaneous let-it-all-go moments on the pond were rare. Sometimes, of course, his sunny mood could cloud up in a hurry, even with me. Like I said, even Mount Reagan needed to blow.

One such occasion came in 1972, after Spiro Agnew had been forced to resign as vice president. It was the first step in the disintegration of the Nixon presidency. Reagan, then governor of California, was a staunch supporter of "Ted" Agnew. He'd gotten to know him when Agnew was governor of Maryland and thought he was a fine man with a weak staff. And he continued to support him even after the bribery charges surfaced and Agnew was forced to retreat in dishonor from public life.

At a staff lunch, I noticed that Reagan was clearly not his jovial self. He was fidgeting with his car keys.

Finally, he blurted out, "Gosh, I talked to Ted Agnew last night, and it's really so sad, so unfair, what happened to him."

I said in front of the entire staff, "Governor, if that guy was Pat Brown, you'd be beating his brains out. We campaigned against that kind of stuff."

He said, "Mike, you're wrong. He's a decent man."

I said, "Look, I'm not arguing that. I'm saying it doesn't help you to defend this man when the world knows he did something unethical."

"You're wrong," he said again, more forcefully this time. "I know Ted Agnew!" I looked up from my notes just in time to see the car keys Reagan had been fidgeting with hurling toward my chest. The crack of them against my sternum left me momentarily short of breath. The man had a Howitzer for an arm.

It wasn't the first time I'd seen Reagan send things flying around the room. There had been some pens, too. Once, he crashed his fist down so hard on the desk that the whole office seemed to shake. These weren't calculated acts. He never would have done that. But when his anger boiled over, the keys and the pens and the fist slamming into the desktop became unconscious points of emphasis. Afterward, inevitably, the person he would be most mad at was himself—enraged that he had lost control.

I remember walking back to his office with him one afternoon after Mount Reagan erupted during a meeting with one of the academic senates from the University of California system. "Oh, my gosh," he kept saying. "That was wrong. I shouldn't have done that."

Anger was one thing; enmity was another. And I don't think Ronald Reagan disliked anybody. I know for sure that there is one person who he loves and respects more than anybody: his wife, Nancy. To capitalize on their closeness and obvious mutual affection, I decided to film a short video for the 1984 Republican National Convention. I came up with an idea to capture a series of facial expressions I wanted from candidate Reagan.

I picked out about fifty photographs that had been taken during the first term and had them all enlarged. Then I had Reagan sit down and give me an on-the-spot reaction as I showed him each photo. The video camera was rolling when I showed him a shot of Nancy carrying a large birthday cake. "Oh, I remember that," he said. "It was taken on my seventieth birthday."

"No," I persisted, "what does it really mean to you?"

He paused and said quietly, "I can't imagine life without her."

I can't imagine them without each other, either. I have heard people say that the Reagan marriage can't really be that happy—that it's some kind of Hollywood fabrication. But the truth is, they really didn't need anybody else; they were inseparable best friends.

Reagan really loathed flying, but he eventually accepted it as a fact of modern life. During one of my first takeoffs with Reagan, I looked across the aisle to catch his lips moving silently. His face was grimaced, and he was looking down as the G-force got stronger.

"Praying for a safe flight?" I joked.

"Actually, I'm saying a prayer for Nancy. If we die today, I'm asking Him to look after her."

He would repeat this exercise every time we flew without her—without fail. That was the real Reagan. He was most at ease when he knew Nancy was nearby, never totally comfortable when she was gone.

On those occasions when Nancy was traveling without him, Reagan would have trouble sleeping. He'd come by my house or that of another close aide for dinner or I'd try to arrange for some fun folks to watch a movie at the White House with him—anything to fill the void.

Once, when I knew he would be spending the weekend without Nancy at Camp David, I suggested that Carolyn and I come up, and bring our daughter, Amanda, and a friend with her. It was Amanda's thirteenth birthday, I told him; maybe the president would like to celebrate it with us. The five of us had a festive dinner that night in the president's lodge, complete with cake and candles. Afterward, Reagan announced that he had a surprise for Amanda, and with that we all settled into comfortable sofas and chairs for a special viewing of Ronald Reagan starring in *Bedtime for Bonzo*, the movie that may be the butt of more jokes and comedy skits than any other film ever made. The chimp who played Bonzo, Reagan explained as the movie went on, was really quite intelligent. It's worth saying again: he was a man without artifice.

Another time, Carolyn and I had been dining in the same lodge when she noticed a set of golf clubs leaning against the wall.

"Is there really a golf course here?" she asked.

"Sure," Reagan answered, walking her out to the deck and showing her the hole that ran just below. "Do you play?"

When Carolyn said that she had once been a pretty decent golfer, Reagan insisted that she take a club with her. He'd have his steward put a bucket of balls out for her in the morning.

The next day, Carolyn was hooking balls into the woods to her left, trying to work the kinks out of a rusty swing, when she heard a voice from above.

"Mind if I try a swing?" the president asked.

After he, too, had hooked a shot, he handed the club back to Carolyn and set off briskly toward the woods.

"Where are you going, sir?" she asked.

"To find the balls," he answered, as if it were the most natural thing in the world for a president to do. Beneath all the pomp and splendor of his office, Reagan really was just a guy named Ron. What's more, he was delighted when he found not just one but three of Carolyn's golf balls.

I, too, have felt the warmth and generosity of the man. In 1975, I had talked Reagan into giving a pro bono speech to the Mzuri Safari Club, a group of conservation-minded hunters based in San Francisco. He delivered a strong speech, and the head of the outfit presented Reagan with a stunning bronze lion as a gift. It was a beautiful work, probably weighing ten pounds or so. I was immediately taken with it and told Reagan as much during the ride home. He studied the piece, probably not seeing the same beauty I saw in it. Within a week, though, the little lion was out of my mind. I never saw it again. Like a lot of such gifts, it simply disappeared, probably into Reagan's attic.

Fast forward to 1985 in the Oval Office: I had given my resignation to the president the night before, and he later asked to see me in private the next morning. As I walked into the Oval Office, there was Reagan standing an inch or two in front of his desk, his hands behind his back.

"Mike, all night I've been trying to think of something to give you that would be a reminder of all the great times we had together."

As he said this, he turned around and picked up the bronze lion from the presidential desk.

"You kinda liked this little thing, as I recall," he said with a beaming smile, his eyes moist. I was totally overcome with emotion, unable to believe that he had remembered after all that time. I took the little lion into my hands and held it tight, and at least metaphorically, I've been holding it tight ever since.

I have an attic stuffed with mementoes—more than an attic, really—but that lion has sat in the place of honor in my den ever since the day Ronald Reagan gave it to me. Every time I look at it, I remember the moment and the man, and the brief peek I was allowed behind the curtain where the real Reagan lives.

It isn't always easy to see, but I know his heart is as big as the full-scale lion my little one was modeled on.

CHAPTER SIX

∽∾

TOUGH TIMES

I've learned that time can be a great healer, my best friend. Time has let me break through the shadows that once surrounded what were to me the toughest times in my relationship with Ronald Reagan. Time has given me the wisdom and perspective to see those days and events more clearly. And time has given me a rare gift, too—a chance to forgive and be forgiven.

The first tough time came to a head on Thanksgiving Day 1979 when I received a call from Nancy asking if I could stop by their home in Pacific Palisades for a quick meeting. It was the autumn before the start of the primary season to select the Republican who would take on Jimmy Carter. We had put the old 1976 team back together, but this time the Easterners—led by John Sears—were determined to get rid of the Reaganites from California. The Sears crew never understood or appreciated the relationship that Reagan had with his California team, and they had no intention of letting us share the ride this time.

Sears had been brought in as the campaign manager in 1976 to help us cover our flank on the East Coast. We were very sensitive to the fact that the Eastern Establishment still dominated elite thinking in America. To them, Reagan's deep conservatism was, if not quite anathema, then at least troubling. Many of them were still smarting from the bitter 1964 convention battle finally won by Barry Goldwater. With our base in California, we thought we needed a pro with connections to the eastern media and politicos. Sears had worked for Nixon and was well regarded by the press, who thought he was a political genius—an image he cultivated by schmoozing the media and keeping them close. Bringing him in gave us instant credibility with this group. Sears hadn't endeared himself to a number of my fellow Californians, but I took full responsibility for bringing Sears and Company back for the 1980 campaign.

Sears, though, simply did not want any competition for Reagan's ear on strategic decision making, and that meant the Californians had to go. He began by driving out Lyn Nofziger, who had been with Reagan since 1966, serving with distinction as his press secretary throughout both terms as governor. Sears made his life miserable, badgering him over the fund-raising effort he was spearheading—a difficult task in the best of circumstances, which these certainly were not. When Lyn had finally had enough, he left quietly, and I mistakenly took over the fund-raising job. I was digging my own grave, and Lyn told me as much. It was only a matter of time, I figured, before Sears and Company made their move—and not much time, either.

I noticed nothing out of the ordinary when

Carolyn dropped me off at the Reagans' that night in November. This all changed when I walked in and Nancy asked if I minded waiting in the bedroom. "The bedroom?" I asked, aghast.

She assured me that it'd be just a few minutes. As I walked by the living room to the bedroom, I saw Reagan huddled with the Easterners, a second sign of sure trouble. I was never excluded from these types of meetings. Fed up after twenty minutes of pacing the floor, I barged out of the bedroom and into the living room in a semi-rage and asked what the hell was going on. Everybody looked at the floor sheepishly. My old friends—yes, we were friends at the time, unified in our desire to see Reagan as president—Sears, Jim Lake, and Charlie Black waited quietly with our candidate.

Reagan spoke up first. "Mike, the fellas here have been telling me about the way you're running the fund-raising efforts, and we're losing money," he said.

Reagan clearly didn't like the position he

found himself in. There was a pained look on his face. He hated this stuff. Here were three guys—newcomers whom he had entrusted to run his campaign—sitting in his living room, trying to purge his longest-serving and most loyal aide. It had come down to this: Deaver or them.

I must admit I was set up nicely. Reagan was right; the fund-raising was not going well at the time. I should have known when I took it over that money would be tight, especially a year before the election with a field crowded with strong contenders like George Bush, Howard Baker, and John Connally. Still, I was stunned and angry, determined not to take this lying down. Surprisingly, calm overtook me just as I was beginning to boil over, and I said coolly, "You need to put somebody in charge, and if these guys are going to sink to these tactics now, what would we do if we ever got to the White House?" Then with Shermanesque resolve I said, "I'm leaving."

I started out of the room, and Reagan followed me. "No, this is not what I want," he said. His pained look became one of utter frustration at the realization that things were spinning out of control. Without breaking my stride for the door—or turning around—I told Reagan, "Sorry, sir, but it's what I want."

Everything came to a head at that moment. I had had enough. I was tired of the dissension, tired of the commute between Los Angeles and Sacramento. I missed my family. All these things were running through my head as I walked out the door that evening. Nor was my mood helped by the fact that I was a major reason why Sears, Lake, and Black were in the Reagan campaign in the first place.

I shut the door, stepped into the California air, and took a deep breath. Then—like someone who has made a glorious exit only to find he has walked into a broom closet—I realized I had no way of getting home. Carolyn had dropped me off; she wouldn't be back to pick me up for several hours. I quietly opened the front door again. Nancy was still alone in the foyer, but I could hear Reagan talking loudly. I put my finger to my lips, seeking Nancy's silence. She complied, and I listened to Reagan admonish the group. "Well, I hope you're happy. The best guy we had just left," he said.

I drew up closer to Nancy. I think she was astonished when I asked not for her behind-the-scenes support to get my job back but for the keys to the Reagan station wagon so I could get home.

I returned to life outside of politics, to my first love—public relations. For several months, I monitored the foundering Reagan campaign with detachment. Reagan just lost the 1980 Iowa caucus, and a reinvigorated George Bush was taking his Big Mo to New Hampshire.

I was spending more time with the family than I ever had. I was able to do some fishing with friends, and I actually started gardening. Life was great, but truth be told, I missed the excitement of being with Reagan. Here he was fighting a bare-knuckles national campaign, and I was planting petunias. It didn't seem right, but I had made up my mind not to go crawling back to a campaign that I missed so much.

In my new role as observer, it was clear to me that Reagan needed his old guys back. It was easy to tell from seeing him on television that he was off his game. After the Iowa debacle, Reagan suggested I stop by his home in Pacific Palisades to catch up. He would be there for a few days licking his wounds. Maybe he needed a distraction from the day-to-day stress of the campaign; maybe he just wanted to see an old friendly face. Regardless, it was great to be with him, and we didn't discuss campaign staffing or the Thanksgiving Day massacre.

Looking back, I think he may have been feeling me out, because a couple of days later, I started getting calls from some of my friends in the campaign saying that Reagan had enough of the "new guard" and was pining to reassemble the California team. Ten days before the critical New Hampshire primary, Ed Meese, a survivor of the purge because of his vast policy expertise, told me Reagan was going to unload

Sears before the primary. He said Reagan decided to do it before the primary no matter where he stood in the polls. Knowing Reagan's distaste for personnel decisions, I told Ed not to hold his breath, but he almost could have.

About two days later I got a call from Nancy. "Would you please come back? Ronnie needs you," she said. I agreed immediately.

My wife, Carolyn, didn't want me to go back. I think this was the first time during our marriage that I had really focused on being a husband and a father. Now Carolyn and the kids would have to share me again with the Reagans.

I rejoined the campaign along with Lyn Nofziger and others who had been pushed out before me. Reagan told the outsiders who had taken over the campaign and driven it into the ground that "things just aren't working out." Later, he said memorably about Sears, "I look him in the eye; and he looks me in the tie." Not only did the boss need straight-shooters, he needed a certain comfort level that the old guard from California brought to the table. For all his supposed campaign smarts, Sears never understood those basic facts about his candidate.

To my surprise, Reagan let the Sears team go the day of the New Hampshire primary, just hours before the voters of the Granite State went for Reagan. Ed was right. He had made up his mind to do it before the votes were counted; win or lose, Reagan was going with his instincts. He was back in charge. If he was going to win this, he'd win it by being himself, something else Sears and Company never cottoned on to. Reagan's candidacy would not rise or fall based on their political genius or how well they lined up vacillating delegates. The key to a Reagan victory would not be found around a green velvet table in a smoke-filled room; it would be found at the campaign rally and the debate lectern and in the television studio where Reagan could look voters in the eye and tell them why he's the right man for the job.

If we Californians had a modicum of smarts, it was that we understood the secret to Reagan's success. He won when people saw him and heard his message. Personal persuasion, not political manipulation, was the secret of Reagan's magic. To Sears, Ronald Reagan was just another political commodity. To us, he was something special, someone we knew the American people would embrace if they got to know him. The campaign and the road to the presidency would now begin in earnest.

Happy as I was to be back in the thick of the fight, I felt a great unease at rejoining the campaign. Although we had spent time together in the interim, Reagan and I had never discussed what happened that evening at his home in Pacific Palisades. I never received assurances it wouldn't happen again, and in truth, I still felt a sense of disappointment that he had let it happen. I didn't say much to Reagan or Nancy that first day back on the campaign plane, hoping instead to get some time alone with them for a heart to heart. I would not wait long.

I was lying on my hotel bed at about nine that night, half dozing over the local news, when a firm knock on the door brought me back to life. It was one of the Reagans' longtime personal aides, Dave Fisher. "The Reagans want to see you, Mike."

In the Reagans' suite, I saw the two of them standing together, still in the clothes they wore at the earlier campaign appearance. Candidate Reagan's tie was off, but he still looked all business with his white shirt and silver cufflinks. He was holding an untouched, fizzing drink in his right hand. This was a most unusual sight, but my confusion was soon cleared up. He nodded politely and held the drink up. "It's scotch and soda, right, stranger?" I nodded and took the drink as we walked toward the living area and took a seat.

Reagan then asked rhetorically and with a half smile, "So what the heck happened to you, anyway?"

I told him a funny thing happened to me last Thanksgiving in Los Angeles, and that "If Mrs. Reagan hadn't given me the keys to the wagon, I would have been out of luck."

Nancy shook her head and looked at me affectionately. "Mike, isn't it about time you started calling me Nancy?"

Although it took me a while to get the Nancy thing down, I knew all was well with the Reagans. This was Reagan's way of clearing the air, only this time his olive branch of choice was a scotch and soda, not a pen and pencil set. I made my point my way; he resolved things his way.

We talked about how great it was to be back together and how we were looking forward to the rest of the race. It felt like old times. They assured me that a rift would never happen again, win or lose. Politics didn't seem that important right then and there, but we all agreed we'd been buffaloed by Sears and that we needed to stick together. As I walked back to my room, I couldn't help but smile as I thought of how close our lives had become.

Those several months were the longest I'd been apart from the Reagans until we drifted apart in 1987.

* * *

During Inaugural Week of January 1985, my body was starting to tell me that I was taking on too much. Not only was I still doing my regular duties as deputy chief of staff to the president, I also served as chairman of the president's second inauguration. I was overworked and exhausted, and the last thing on my mind was my health.

There was also a more insidious and related problem to deal with—my abuse of alcohol. This was a part of my life that I kept to myself and constantly denied even as I watched it eat away at my physical and spiritual well-being. Both my parents fought and suffered from the disease throughout my childhood, although both were ultimately able to abstain from drinking for their last twenty-five or thirty years. Maybe I thought that the same thing would happen to me without my having to make any hard choices, that there would come a magic time when alcohol would just disappear from my life. Whether it's booze, drugs, or whatever, addicts are capable of tortured logic.

I had controlled "it" as best I could while in

the White House, but sometime in the last six months there my "filter" broke. Booze became my master, not my servant. During a routine checkup, the Navy doctors noticed my blood pressure was elevated and put me on a regimen of beta blockers. For the next several days, I kept plugging along like nothing was different. But the combination of stress, alcohol, and beta blockers is a formula for disaster, and pretty soon people began to take notice.

Everybody from my wife to Chief of Staff Jim Baker told me I looked like a corpse. I finally agreed to go to the hospital after somebody pointed out that the left side of my face was sagging dramatically. I would spend the next ten days in Georgetown Hospital and another week at home recovering from kidney failure and its impact on my neurological system. Pinned down in that hospital bed and later recovering at home, I realized that it was time to get out.

On the day I decided to pull the trigger, I called Nancy and told her we needed to talk. I was more worried about telling her because

I knew she'd want to talk it out, but I was surprised at how keenly she understood my predicament. She was supportive and very understanding of my need to move on. I then went down to the Oval Office to tell the president. I knew that his only concern would be whether this was what I really wanted to do. I told him as much, and there was no more discussion. He just wanted to make sure that I was happy and comfortable. To help ease the transition of the second administration and the president's upcoming trip to Bitburg, Germany, I agreed to stay on through mid-May.

Late that spring, I jumped into my own business, Michael Deaver and Associates, hoping and pretending that everything was fine. After all, I told myself, it wasn't the drinking that had been the problem. It was the long hours I'd put in at the White House, arriving by 6:30 most mornings and often not leaving until 9:00 or 10:00 at night. But there comes a moment of truth for every alcoholic. He can either change his life or eventually ruin himself. This often depends on the loved ones around him who have the power to confront and guide. I was one of the lucky ones. My family essentially took control, telling me the drinking was over and that I'd be getting help.

I learned a remarkable truth: You gain by giving up. I left my fledgling business in the hands of my colleagues and checked into Father Martin's clinic in Maryland. At forty-seven years of age, I had to start all over. I had to rebuild my life, my marriage, my livelihood, and my relationships.

Once I was back on my feet, business at Michael Deaver and Associates was good, perhaps too good. Blue-chip clients were literally breaking down my door, anxious to take advantage of my position with the Reagans. Unfortunately for me, I had abandoned all I learned at Reagan's side. I so let success go to my head that I even allowed *Time* magazine to do a cover story on me, complete with a soon-to-be-famous cover photo that showed me stepping out of a limousine. When the "pro-

~~~

file" turned out to be a hit piece on Washington corruption, I suddenly found myself the poster boy for a system gone awry. I should have kept my head down like I've done my entire life, staying behind the scenes where I am most comfortable. But I was out there too far and even Nancy noticed it. She called the day the *Time* piece came out. "Mike, I think you made a big mistake," she said. "I think you're going to regret posing for that picture." At first I thought Nancy was being too paranoid, but the more I thought about it, the more I remembered how good her antenna was at picking up things like this.

I would soon find out how right she was. In Washington, if you ever look like you're having too much fun or making too much money, somebody will kick you while you're up. The Washington elite media that had begged for access while I was at the White House trashed me as someone wantonly cashing in on his ties with a sitting president. As usual, they overdid it, and I found myself in a hell of a mess.

The president stood by me when asked.

"Mike hasn't done anything wrong. He has never put the arm on me," he said. Still, it was a horrendous feeling, knowing that Reagan felt a need to defend me. For obvious reasons, I cut back my contacts with the White House.

Eventually a special prosecutor was appointed to investigate whether I had violated the lobbying restrictions imposed on former senior White House aides. I won't get into the merits of the case here. Suffice it to say that the painful legal process came to an end.

Nancy would send her wishes through emissaries, rightly cautious of being tarred by Washington's pariah of the month. While Democratic titan Bob Strauss called me religiously every Friday night to see how I was coping with the rigors of the investigation and later the trial, few of the old Reaganites bothered. One exception was Richard Helms, the former CIA director, who would kindly insist on taking me to lunch regularly at Washington's best places. The legal battle left me penniless. I felt alone. Flat on my back, I called my old political mentor Stu Spencer for advice on where I could find fifty thousand dollars to make a tax payment. "I'll have a check in two days," he said. I fought back because that was the last thing I expected, but I took his money and repaid him in six months, eternally grateful.

Getting back on my feet financially would take hard work. Getting a sense of balance back in my life would prove, in some ways, harder still. Not only was I broke and embarrassed, I kept casting around in my own mind for people to blame my situation on, and as I did that, I kept coming to my old friends, Ronald and Nancy Reagan. How could they seem to be getting along so well without me? Why was it that I heard from them only secondhand these days? How hard would it be for Ronald Reagan to pick up the phone and call me? Hadn't I given him the best years of my life? I wanted desperately to move on, to put the past behind me, but my self-pity and growing bitterness just wouldn't let me.

It was in such a state of mind that I found myself alone in the Century Plaza Hotel on February 6, 1991. I had flown out to Los Angeles to try to scare up some new business on the West Coast, but try as I might to concentrate on the contacts I would be making, all I could think about was my life with Ronald Reagan and what he meant to me. And that's when it finally dawned on me: if I was ever to move forward again, I'd have to revisit the past one more time and ask his forgiveness.

It had been three years since I'd seen Ronald or Nancy Reagan, and longer than that since I had talked with them. I decided then and there to reach the Reagans. I picked up my address book and dialed their home number.

While I shouldn't have been, I was surprised to hear Nancy's pleasant voice answer. I almost hung up. "Hello," she said again.

I told her I was in Los Angeles and that I'd like to stop by in the next few days and say hello.

She sounded delighted, "Of course, but why don't you come now?" she asked. "We're both here."

I agreed but needed to know exactly where "here" was. I had no idea where their new home stood, so Nancy dutifully gave me a crisp set of directions.

I felt apprehensive as I drove out to Bel Air. I could barely focus on the directions but did manage to stumble upon the place the Reagans had called home since leaving the White House three years earlier. It was larger than their old house in Pacific Palisades, with a contemporary California feel to it and a commanding view of the Los Angeles basin. I still wasn't sure I'd made the right decision in calling them. Perhaps I should go back and write a letter, I thought. Maybe I wasn't ready for a face-to-face meeting. I made some small talk with the Secret Service agent near the front door, hoping to delay my entrance.

Nancy would have none of it, though, as she opened the door and literally pulled me in. After giving me a hug, she briskly walked me to the den, where her husband was relaxing. The Reagans could not have been warmer. It was clear they were genuinely glad to see me. He stood up and shook my hand tightly. They both looked wonderful, and it felt like old times as they took me on a condensed tour of their home. We then went back into the den where Reagan poured us some tall glasses of iced tea. There was a lull in the conversation as I nervously looked around the room.

Finally I told them why I was there. "It is important to me to say that I am really very sorry for some of the things that happened over the last few years," I began. I told them how I wished my trial had never happened. I told them I was sorry for it and sorry for what had been said over the airwaves and in print. The press had twisted the trial so badly, I wondered if the image of Mike Deaver on the evening news would tar my relationships with the Reagans. I knew that much of this had been painful for them, too.

They looked at each other and smiled. Nancy leaned over and took my hand, "Mike, forget all that," she said, "we're just glad you're back with us." We talked for a little longer until a quiet moment settled over the conversation and I stood up to break the silence. I needed to get out of that house then and there.

Reagan announced, to no one in particular, that he'd walk me to the door. I was wondering if he wanted to get this over with as badly as I did. As he opened the door, he motioned me outside onto the flagstone landing. Stepping out, I heard the door click quietly shut behind me and wondered for a moment if he had given me the slip. But no, he was still standing next to me. We stayed that way, silent, for what seemed a full minute or two.

"Mike, today's my eightieth birthday, and this is the best birthday present I could ask for."

He pulled me in and gave me a very uncharacteristic hug. I was overcome by emotion. Not just because of Reagan's kindness, but the fact that all this was happening on his birthday, an event that hadn't even crossed my mind. February 6, how could I have forgotten?

We stood together for a moment, then he

walked me to my car. Pulling away, I watched in the rearview mirror as he retreated to his home. I tried to remember a similar show of emotion by the man and came up blank. Reagan was simply not the hugging type. He didn't even like shaking hands much.

I smiled to myself. Driving out to the Reagans' house, I had felt sluggish, burdened by my own weakness; now, I felt something approaching a pure euphoria. The monkey was off my back; the war with myself was over. If I never saw Ronald Reagan again, I would have this moment as a precious keepsake.

About a month later I received an unsolicited call from Reagan asking me to consider managing the opening event for his new library. I agreed immediately. Our friendship was no longer on life support—I could put away the defibrillators.

It was great working with the Gipper and Nancy again. I wanted the library opening to be a memorable Reagan event and it was. For the first time in the history of the republic, five former presidents gathered at the same location. It marked a new chapter in my relationship with the Reagans.

I had doubted Reagan's affection for me during our silent period from 1987 to 1991, and it was difficult for me to fully understand. In time, though, I came to realize that Reagan is so totally complete in himself that the only person he really needs is Nancy. Yes, I am a very special person in his life, but if I am out of sight for a few weeks or years, that's okay, too. It's Nancy that he wants and needs to be around all the time.

I came to see, too, that Ronald Reagan thought a choice had been made to keep our lives apart during that silent period and that I had made it—and maybe, in fact, I had without ever realizing it. I often heard during those periods of separation from friends and others who were with him when he talked about me, leaving the impression that he was still thinking about me.

One recent evening Nancy called me at home to tell me that he asked her, "Nancy, where is Mike Deaver?" I like believing that even today, somewhere in the clouds that darken his memory, there is a flash once in a while of memories of the good times we shared.

CHAPTER 7

THE LONGEST GOOD-BYE

Not long after I'd gone to work for Governor Ronald Reagan, he asked me to sit in on a meeting with some state assemblymen. I arrived at his office a few minutes early, and he directed me to take a seat across from his desk. Reagan had just started across the room for a glass of water when his secretary announced that his wife was on the line. He spun on his heel and lunged at the phone. As he sat in his chair, he greeted her with a slow, drawn out "Hello." I stood up to leave, thinking this was a private moment, but he waved me back into my seat, raising his index finger to indicate this would be just a short call.

I reviewed my notes as they talked, looking up only when I heard the governor first say "bye." He was sitting erect in his chair. He listened for a second more, leaned slightly closer to the phone's cradle, and uttered another, softer "bye." Five or six more "byes" followed until Reagan's head was literally by the phone cradle, the receiver still to his ear. There was a last "bye," no louder than a puff of wind, before he hung the phone up gently, as if it were made of fine crystal. Then he looked up at me and winked.

I opened my mouth, but he anticipated my question. "It's funny, but neither of us ever wants to be the last one to say 'good-bye.'"

Now, of course, that choice has been taken from him. As he begins his ninetieth year, Ronald Reagan should be basking in the warmth and love of a thankful nation. The Evil Empire is gone. Democratic capitalism is triumphing over totalitarianism all across the globe. America really is what Reagan always insisted it was: that "shining city on the hill," the "last great hope of mankind."

Reagan was right all along, and he should get to enjoy how right he was. But the sunset he has ridden off into is one without colors. It's a place where birds don't sing, where there are no sweeping vistas, where moment follows moment in a dull tedium of existence. For a man who loved life so much, it's a cruel, cruel end.

I first realized something was wrong with Ronald Reagan when I was pressing him to retell a story he'd told hundreds of times.

The White House days were over. We were in his Los Angeles office, meeting with representatives of a production company who were considering doing a movie about the late Lee Atwater, the Republican national chairman who had died several years earlier of a brain tumor. The movie people had asked for an opportunity to meet the president, and Reagan had agreed. He looked great and seemed to be in good spirits, but I soon realized things were not the same.

It started with an innocent question from one of the guests. What was the secret of Reagan's

∽⌒∾

indefatigable optimism? He was amazed with the way Reagan kept his good humor regardless of circumstances and wanted to know how he did it. For Reagan, this was a pure softball, right over the center of the plate. He always had a supply of stories ready by way of illustration, and the one he saved for telling about his sunny view of life was among his best. In a nutshell, it went like this:

A young couple has a pair of twin boys, both seven years old. One boy is a hopeless optimist, the other a brooding pessimist. A shrink is brought in to conduct a thorough analysis. He takes the pessimist to a room filled with the latest and greatest toys. When he opens the door and tells the child that all the toys are his, the boy begins to sob uncontrollably. What's wrong, he asks. "These toys will surely break someday," the boy answers. "Then what'll I do?" Next, the optimistic twin is taken to a decaying barn. The psychiatrist opens the barn doors, revealing a two-story pile of fresh, steaming manure. Undeterred, the kid grabs a shovel and starts digging away with a mile-wide smile on his face. The doctor can't believe it and asks what he's doing. "Well, with all this manure," the boy says, still smiling, "I figure there's gotta be a pony in here somewhere."

The story never fails to get a laugh. Even if you've heard the joke a thousand times from other people, Reagan's delivery and the twinkle in his eyes did more than humor you; it helped you actually feel his sense of optimism. To me, Reagan was always that little kid shoveling. I wanted the guests that day to see that side of Reagan, but instead of launching into the tale, he just looked down at the floor. "Well, I guess it's just something I've always had," he finally said. "I try never to be a pessimist."

Startled, I urged him on.

"Mr. President, tell him the one about the pony."

"Which one's that?" he said.

"Oh, come on," I said. I thought he was pulling my leg. "The one about the optimist and the pessimist, the one you've told a million times."

The frozen look on Ronald Reagan's face

told me something was wrong. He was obviously embarrassed. I tried to save the moment by telling the story myself. Reagan laughed with the rest of the room, but it wasn't a laugh of complicity. He laughed as if he had never heard the parable before. I was stunned.

When the movie guests were gone, I went to Reagan's assistant, Fred Ryan, and told him what had happened. Ryan wasn't surprised. These types of incidents were happening more and more frequently, he said. There were even occasions when old friends and former aides would stop by and Reagan didn't recognize them.

Deep down, I thought it was just old age catching up with him, nothing more. Nancy tells me she had the same first reaction to his forgetfulness. He was getting old; it's natural to have these memory lapses. Neither of us even thought of Alzheimer's. It was something that happened to other people, not to Ronald Reagan.

In preparing for this book, I sat down with Nancy Reagan the week before the August

2000 Democratic National Convention in Los Angeles to relive some of our shared memories. We met at one of her favorite places for lunch, the Bel-Air Hotel. Over the last five years, Nancy has shared with me the difficulties of their situation, but I feel honor bound not to betray that trust and write about those times. She reminded me of a toast he had offered to her and other family members on Christmas 1980, on the eve of their departure for Washington and the White House: "This toast is for all of us. Not for what we're about to become, but for what we've been, to each other, for so many years." That's how she wants Ronnie remembered: not for what he has become and is becoming, but for what he was for so many years.

Over lunch, we mostly talked about those days and the good memories, but it was impossible to completely ignore the reality of the moment. We were interrupted by well-wishers at least five times as we ate. All of them wanted to know the same thing that I did: "How's our president?" Nancy would demur each time, simply saying that he's getting along.

She still can't believe that, after all Ronald Reagan has been through, Alzheimer's disease is what brought the great man down. "My goodness, he's been shot, he's had cancer twice and suffered a terrible riding accident," she said. Nancy firmly believes that a serious riding accident in 1989 contributed to the onset of the disease: "That set the stage for it."

When Reagan was diagnosed with Alzheimer's at the Mayo Clinic in the summer of 1994, Nancy said she was initially angry. That quickly faded into denial, then reluctantly into acceptance. Nancy had decided early on that he would not be going to any home or hospital specializing in caring for those afflicted with Alzheimer's. She'd be carrying the load herself. There is no bitterness, only an overwhelming sense of duty and responsibility.

One of my goals in talking with Nancy was to learn more about the circumstances surrounding Reagan's famous final letter to the American people. The diagnosis had come in the summer of 1994. Several months later, on November 5, 1994, Reagan took his wife by the hand, she says, and led her into their study. She sat across from him silently as he grasped his trusty fountain pen and choreographed his final exit from the public stage:

My Fellow Americans,

I have recently been told that I am one of the millions of Americans who will be afflicted with Alzheimer's Disease.

Upon learning this news, Nancy and I had to decide whether as private citizens we would keep this a private matter or whether we would make this news known in a public way.

In the past Nancy suffered from breast cancer and I had my cancer surgeries. We found through our open disclosures we were able to raise public awareness. We were happy that as a result many more people underwent testing. They were treated in early stages and able to return to normal, healthy lives.

~∾

So now, we feel it is important to share it with you. In opening our hearts, we hope this might promote a greater awareness of this condition. Perhaps it will encourage a clearer understanding of the individuals and families who are affected by it.

At the moment I feel just fine. I intend to live the remainder of the years God gives me on this earth doing the things I have always done. I will continue to share life's journey with my beloved Nancy and my family. I plan to enjoy the great outdoors and stay in touch with my friends and supporters.

Unfortunately, as Alzheimer's Disease progresses, the family often bears a heavy burden. I only wish there was some way I could spare Nancy from this painful experience. When the time comes, I am confident that with your help she will face it with faith and courage.

In closing let me thank you, the American people, for giving me the great honor of allowing me to serve as your president. When the Lord calls me home, whenever that may be, I will leave with the greatest love for this country of ours and eternal optimism for its future.

I now begin the journey that will lead me into the sunset of my life. I know that for America there will always be a bright dawn ahead.

Thank you my friends. May God always bless you.

Sincerely,
Ronald Reagan

He read the letter once to himself, capped his pen, and slid the letter to Nancy. There had been no discussion about how the letter would be worded, no exhaustive kitchen table chats about how best to let America know. There had just been these few minutes of composition, and this letter that is inseparable, heart and soul, from the man who wrote it.

Once again, Reagan's timing was impeccable. It was as if Providence had given him a small coherent window in which to make his intentions known. Although physically he felt fine at the time, Reagan always knew when to step off the stage. The time was right. Nancy agreed, and the letter was soon made public.

As in everything Reagan did, the letter was also about much more than the man himself. In excess of 4 million Americans suffer from Alzheimer's, with another 400,000 diagnosed every year. When the baby-boomer generation begins to reach its age of greatest vulnerability later this decade, experts predict that more than 14 million of us will be afflicted, but it's not just the sick who must pay the price. As Reagan's letter suggests, Alzheimer's exacts a horrible toll on families, too.

The Reagans—especially Nancy and the late Maureen Reagan, the president's daughter—have sought to use his condition as a way to promote awareness and seek additional funding for research into Alzheimer's and eventually a cure. Since Reagan penned his open letter in 1994, Maureen told me that more and more caregivers share information about how the disease has decimated their families and how they've learned to cope with the burden.

For all the public good the Reagans have managed to do, though, Alzheimer's ultimately comes down to a very private pain. Nancy girded herself for the tough times ahead, she said, but nothing could really prepare her for what was coming. She would try to keep her husband happy, taking him to his beloved ranch several times, but it seemed to do more harm than good. "He just didn't want to be there," she said.

Soon other old places and pleasures lost their allure, and the shared memories that were so dear to them became blurs and, finally, vacuums. There would be no more conversations starting, "Honey, remember when . . . ?"

Nancy remains stoic. "When we said 'for better or worse,' we meant it," she said. She retains much of the Gipper's unflagging optimism. "Everything happens for a reason," she would tell me twice over lunch. Her strength, always great, is little short of amazing these days. Despite a bout of double pneumonia that struck her down over Christmas 1999, Nancy continues to field calls from Republican leaders and to monitor national politics closely. When I talked with her recently, she had just spent the day at the Reagan Library signing fifteen hundred copies of her own book. But the beautiful dining room at Bel-Air is no longer Nancy's natural setting, and outings to anywhere are far too rare. There's a sick husband to look after, a husband who still knows her even when so much else has been taken from him. And sometimes, despite all the strength and optimism in the world, the reasons for which things happen can be very hard to find.

The day would come when I would see Ronald Wilson Reagan for the last time in my life. His Alzheimer's had ratcheted up its already tight grip on the president's mind. Nancy had let it be known that she was soon going to restrict visitors to immediate family, so I needed to schedule my final good-bye.

In a way, I was almost relieved it would be my last time seeing Reagan. Unlike Nancy, I don't think I'm strong enough to bear such close witness to the man's decline.

By ending visits, Nancy was clearly acting in her husband's best interests, but I think in the back of her mind she was also protecting his friends. They shouldn't have to see him at his worst. She would not let the good times be elbowed out of the way by this invisible thief of memories.

There was a lot less activity in Reagan's office on this day. When I had last visited, there had been at least a semblance of business. Maybe it was just busy work, but at least the phones were ringing. On this day, there was an eerie silence.

I waited for Joanne Drake to escort me into his office, but she was nowhere to be found. To pass the time, I made small talk about the traffic with the Secret Service agent in charge, but he quickly became bored. Why didn't I just go in on my own, he suggested.

As I pushed the door open, the knob for some reason slipped out of my hand, and the door thumped loudly against the stop. Ronald Reagan was sitting at his desk, reading a large book. Despite the noise, he didn't look up.

Surprised, I closed the door slowly and quietly, and just stood there for about fifteen seconds or so. He still didn't look up.

My eyes traveled around the office, noting changes. I wondered if Nancy had slowly begun removing things as the disease made its progress. There were far fewer paintings on the wall. Small stacks of books lined the room on the floor. It looked like he was preparing for a move. Certainly, this was no longer the Oval Office West.

I refocused my gaze on Reagan. He looked pretty good, I thought. Blue suit, French cuffs—for a man then in his late eighties, he was well turned-out. Some things never change, thank God. Finally, though, I realized that I could stand there all day and he wouldn't notice me. I'd have to take charge if there was going to be any conversation.

"Hi," I blurted out too loudly as I moved awkwardly toward his desk. I began shifting the leather datebook I habitually carry to my left hand, preparing to grasp once again the firm hand of my old friend and boss.

Only his head moved as he finally looked up. His gaze, so questioning and unrecognizing, was new to me. "Yes," he said. His voice was polite, as always, but he spoke with a tone I had never heard from him before. His hands stayed on his book.

With that, he resumed his reading and said nothing more. I stood there confused and saddened. He had no idea who I was, but I wasn't giving up yet. Quietly placing my datebook on the corner of his otherwise barren desk, I sat in a chair to his left and then pulled it closer to him, determined to get his attention.

"Whatcha reading?" I asked jovially.

For years, as Reagan and I crisscrossed the continent together, reading had been part of our routine. I'd settle into my seat on the plane and crack open some *New York Times* bestselling novel. After getting his sack of peanuts and icy Coca-Cola, Reagan would begin thumbing through the reams of memos and briefing documents that made up his "must read" file. Then, almost always, he'd set his papers briefly aside and, with a hint of envy and curiosity, ask me about my book.

Now, that time suddenly seemed so long ago. There was a pained look on Reagan's face as he answered my question. "A book" was all he said, and he looked back down.

I sat still for maybe three minutes as it dawned on me that I'd gone about this visit all wrong. I'd bulldozed in on him, trying to act like the old friend I was, but he saw only an intruder. I stood and moved slowly next to him, and as I did so, he looked up and gave me a half smile. Grateful for even a modicum of his confidence, I walked behind him and watched his busy fingers move from left to right on the large yellowing pages.

"What book?" I asked, finally focusing on it.

"A horse book," he said quietly.

For the first time, I looked at what he was holding in his hands and realized it was a picture book about Traveller, General Robert E. Lee's horse. I felt like crying.

I'd been told that when his staff cleaned out Reagan's old desk in his Century City office, after it became apparent that he wouldn't be leaving home anymore, they came upon a well-thumbed, time-worn letter from his mother in the top drawer. The content was nothing special—Nelle had sent it when he first moved to California decades ago—but clearly he had kept it close at hand as a reminder of a mother's love. Watching him now, as his finger worked laboriously across the page, I remembered the advice Nelle had given him years and years ago: "If you learn to love reading, you will never be alone."

Today, I was the lonely one. I must have said something as I turned and left the office—"It's wonderful seeing you," maybe. "I've got to go now." But I no longer remember what it was, and the man I had come to see wouldn't have heard it in any event. Ronald Reagan had heeded his mother and lost himself in a peaceful, solitary place.

If Ronald Reagan's life were a movie, it would fade to credits now, the houselights would come up, and we could all dry our tears and go home. But Reagan was always far more than an actor, and no life that has intertwined itself so richly and deeply with history can be judged only by its end.

Just as Reagan belonged to the American people, so the American people in a very real sense belonged to him. He helped to bring out the best in us, in our nation; and this nation and its people helped to bring out the best in him. His was an American life through and through, a promise that the system works, that any kid can grow up to be president. And if Ronald Reagan is playing this final act alone, he still has the one person beside him who he most wants and needs to be there, the one person who will never leave him.

At night, Nancy told me, Reagan still reaches out in silence to see if his beloved wife is next to him. She always is.

May 10, 1985

Dear Mike:

You know I've accepted your resignation orally,
but I suppose I have to put something down on
paper -- after all, this is Washington. The
only place I haven't accepted it is in my heart,
and there I never will.

I've come to the conclusion that Nancy and I
will both agree you will bodily leave the West
Wing. You will no longer bear a government
title. You will not actually handle such things
as schedule, trips, etc., but that's as far as
we go. You will continue to be a part of our
lives. We will have concern -- one for the other,
and refuse surgery that would in any way remove
you from a relationship that is part of our life-
support system. In return, we will continue to
be eternally grateful.

Sincerely, Ron

The Honorable Michael K. Deaver
Assistant to the President
    and Deputy Chief of Staff
The White House
Washington, D.C. 20500

Acknowledgments

The author is grateful to the many people who helped make this book possible in so many ways.

First and foremost is Jeff Surrell, who over the past year has written and researched for me, and has made the project a reality. I thank Rob Rehg and Howard Means for lending their prose, polish, and insight. My editor at Harper-Collins, Mauro DiPreta, and my old friend Bill Adler gave me encouragement and counsel throughout the process.

I thank the Edelman family for believing in me.

I am grateful to Stu Spencer, Lyn Nofziger,

Acknowledgments

and Joanne Drake for helping me remember. More important, though, is their selfless contributions to Ronald Reagan himself.

I will never forget Nancy Reagan's patience and suggestions, and her willingness to spend time even when her days were full caring for her husband.

In addition to all of the above, I've been blessed with many friends and supporters who have continued to be a source of great comfort and strength.

At the risk of offending some, I thank Pat Jacobsen of Fort Worth, Texas, who through thick and thin remained steadfast in her devotion to me and my family.

Most of all, I want to thank Ronald Reagan for offering me a front-row seat in the ride of a lifetime.

LESSONS DRAWN FROM THE LIFE AND EXAMPLE OF RONALD REAGAN

How RONALD REAGAN CHANGED MY LIFE

by Peter Robinson

hardcover: 0-06-05239-9 • $24.95 US • $38.95 Can
trade paperback (coming soon) : 0-06-052400-6 • $14.95 US • $22.95 Can
audio cassette : 0-06-055633-1• $25.95 US • $39.95 Can
audio CD : 0-06-055634-X• $29.95 US • $45.95 Can
large print : 0-06-055814-8• $24.95 US • $38.95 Can

IN 1982, PETER ROBINSON WAS HIRED AS A SPEECHWRITER in the Reagan White House where the example Reagan set— as a confident, principled, generous-spirited older man— inspired those around him. At the core of *How Ronald Reagan Changed My Life* are ten lessons Robinson learned from Reagan, yet it is also an account of one man's profound respect and affection for the president who changed his life.

"Engaging."
Wall Street Journal

"The best portrait of Reagan this side of the Reagan library."
William F. Buckley Jr.

An Insider's View
of
Ron's Beloved Nancy

NANCY
A Portrait of My Years with Nancy Reagan
BY MICHAEL K. DEAVER

hardcover: 0-06-008739-0 • $24.95 US • $38.95 Can
audio cassette: 0-06-058529-3 • $25.95 US • $39.95 Can
audio CD : 0-06-058530-7 • $29.95 US • $45.95 Can
large print : 0-06-058978-7 • $24.95 US • $38.95 Can

I**N THIS THOUGHTFUL FOLLOW-UP** to *A Different Drummer*, Deaver continues the behind-the-scenes story of Ronald and Nancy Reagan, focusing on the time he spent with the First Lady.

Informative and engaging, *Nancy* is a loving tribute to the woman who has become the epitome of temperance, refinement, elegance, and most important, allegiance and fidelity.

**THE STORY OF
RONALD REAGAN'S SPIRITUALITY—
AND HOW IT FOREVER CHANGED THE WORLD**

GOD
and

RONALD REAGAN

a spiritual life

by
PAUL KENGOR

0-06-057141-1 • $26.95 US • $41.95 Can

AN INTENSELY PRIVATE MAN, Ronald Reagan kept his personal religious beliefs relatively quiet over the years. But as presidential historian Paul Kengor shows, Reagan's own words demonstrate that his religious orientation was shaped in his childhood—and was retained with extraordinary consistency throughout his life.

**"He who introduces into public office the
principles of primitive Christianity will
change the face of the world."**
Ronald Reagan, 1967, quoting Benjamin Franklin

The Wit and Wisdom of
Ronald Reagan
One of the Most Beloved
Presidents in Modern American History

RONALD REAGAN
The Great Communicator
Introduction by Nancy Reagan
Afterword by Peggy Noonan

Edited by Frederick J. Ryan, Jr.
0-06-093350-X • $20.00 US • $29.95 Can

THROUGH MORE THAN HALF A CENTURY OF PUBLIC LIFE, Ronald Reagan spoke with consistency and contagious optimism to the hearts and minds of the American people. This unique collection of memorable quotations by the former President is a spirited tribute to one of this century's greatest personalities and political leaders.